Praise for Penney

frequency

"Penney Peirce has crafted a highly useful book that will help you reliably recognize the core vibration of your truest self and harness your sensitivity— so all areas of your life can change for the better."

Carol Adrienne, PhD, author of *The Purpose of Your Life*

"Penney Peirce, a master teacher, has gifted us with this powerful book that weaves so many strands in the consciousness tapestry into a coherent whole."

Marcia Emery, PhD, author of *PowerHunch!* and *The Intuitive Healer*

"*Frequency* is an amazing 'handbook for the future,' synthesizing diverse views for expanding our perception and developing extraordinary human capacities."

Hal Zina Bennett, author of *The Lens of Perception*

"From the title onward, Penney Peirce captures and explains the often misunderstood and overlooked subtleties of energy and vibration. As a psychic detective, I use body sensitivity and the ability to discern subtle energy for crime scene work, tuning in to the frequencies of a location to re-create what took place there. I've also taught children and families how to master their sensitivity to vibrations to ensure personal safety. *Frequency* offers readers the tools to recognize and develop these sensitivities within themselves to speed personal growth and find greater freedom in life."

Pam Coronado, intuitive investigator and costar of
Sensing Murder, Discovery Channel

"*Frequency* reveals the simplicity that underlies apparent chaos. In detailing and giving examples of how to get in touch with the soul's purpose, action, and connection to others, Penney Peirce brings us the gift of how to live in alignment with the magnificence of who we truly are. This contribution rises as seminal in the transformation process for every person."

Joan C. King, PhD, neuroscientist, professor emerita,
Tufts University School of Medicine and author of *Cellular Wisdom*

"This book takes *The Secret* to yet another level. At the center of it is an important truth. The left brain will argue with it, but if you listen to what Penney Peirce is saying with an ear attuned to what resonates in you, you will gain something lasting that really matters."

Don Joseph Goewey, author of *Mystic Cool*

"*Frequency* is filled with Penney's profound wisdom and is deeply helpful to anyone wanting to raise their frequency, achieve inner/outer unity, and move toward emotional enlightenment."

Margaret Paul, PhD, author of *Inner Bonding,* and coauthor of *Do I Have to Give Up Me to Be Loved By You?* and *Healing Your Aloneness*

"A lot has been said about frequency and raising vibration in general, but this book is a well thought out and specific road map to the new realms of consciousness that we are all entering."

Hope and Randy Mead, creators of the movie *Orbs: The Veil Is Lifting*

"We are all affected by positive and negative energy whether we realize it or not. With the wisdom contained in this book, you will learn how to raise the level of your vibration—your frequency—to benefit yourself and humankind in miraculous ways."

Masaru Emoto, author of *The Hidden Messages in Water*

"Many indicators tell us we are about to experience a rapid transition to a new world that will change the essential nature to who we are and how we understand reality. Transcending the coming chaos is possible with the toolset provided here. *Frequency* is an unprecedented gift for the person who is ready to evolve."

John L. Peterson, founder of the Arlington Institute and author of *A Vision for 2012*

"Seeing ourselves as energy beings is the most important breakthrough of our times. In *Frequency*, Penney Peirce clarifies many of the new energy principles that have previously been unacknowledged, but which we can now intentionally use to keep ourselves healthy and improve the realities we live in. I laughed out loud when I read this book, and enjoyed it immensely."

Richard Bartlett, author of *Matrix Energetics* and *The Physics of Miracles*

frequency

the Power of Personal Vibration

Penney Peirce

FOREWORD BY MICHAEL BERNARD BECKWITH

ATRIA PAPERBACK
New York London Toronto Sydney New Delhi

BEYOND WORDS
Portland, Oregon

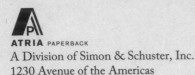

ATRIA PAPERBACK
A Division of Simon & Schuster, Inc.
1230 Avenue of the Americas
New York, NY 10020

BEYOND WORDS
1750 S.W. Skyline Blvd., Suite 20
Portland, OR 97221-2543
503-531-8700 / 503-531-8773 fax
www.beyondword.com

Managing editor: Lindsay S. Brown
Editor: Julie Steigerwaldt
Copyeditor: Meadowlark Publishing Services
Proofreader: Henry Covey, Cogitate Editing
Design: Devon Smith
Composition: William H. Brunson Typography Services

First Atria Books/Beyond Words trade paperback edition August 2011

For more information about special discounts for bulk purchases, please contact Simon & Schuster Special Sales at 1-866-506-1949 or business@simonandschuster.com.

The Simon & Schuster Speakers Bureau can bring authors to your live event. For more information or to book an event, contact the Simon & Schuster Speakers Bureau at 1-866-248-3049 or visit our website at www.simonspeakers.com.

Manufactured in the United States of America

40 39 38 37 36 35 34

The Library of Congress has cataloged this hardcover edition as follows:

Peirce, Penney.
Frequency : the power of personal vibration / Penney Peirce.
    p. cm.
    1. Vibration—Miscellanea. 2. Self-realization—Miscellanea. I. Title.
BF1999.P417 2009
131--dc22

                                    2008044312

ISBN: 978-1-58270-212-4 (hc)
ISBN: 978-1-58270-215-5 (pkb)
ISBN: 978-1-4165-4637-5 (eBook)

The corporate mission of Beyond Words Publishing, Inc.: *Inspire to Integrity*

For the people who experience the most painful suffering,
who feel trapped and hopeless,
who feel life makes no sense and seems utterly cruel—
there is a place in the center of each moment,
available right now,
where freedom and compassionate truth wait
to welcome all of us home.

Contents

Acknowledgments

A handful of people have been especially helpful to me in the creation of this book. Rod McDaniel contributed several important terms, such as "home frequency," that dramatically and immediately aligned and opened me to new insights and information. I'm also grateful for his clarifying feedback on the manuscript and some of his beautiful, original translations of Rainer Maria Rilke's work. Others patiently read the first draft and gave me great insights as well. Much appreciation goes to Darryl Lundahl, Cameron Hogan, Pam Kramer, Henry Smiley, Anne Lewis, Barbara Haury, Anthony Wright, Jim White, and Joan Charles. I am also deeply thankful to my mother, Skip Eby, and my sister, Paula Peirce, for their interest, reality checks, and loyal support, even if they're not always sure what I'm talking about. Thanks, too, to Chris Lenz and Karen Malik from The Monroe Institute for their precise input.

Working with Cynthia Black and the team at Beyond Words Publishing has been a sweet experience. They took a chance on me long ago, publishing *The Intuitive Way*, and I am happy to be back with them. They are sophisticated and warm, and Cindy has been a real support as I have moved into this new territory. Marie Hix has been upbeat and unfailingly helpful, weaving the threads of the project together. I couldn't have asked for a more talented, knowledgeable, and considerate editor than Julie Steigerwaldt. Thanks, too, to Lindsay Brown, Sara Blum, Devon Smith, and Bill Brunson.

Finally, I am grateful to Michael Bernard Beckwith for taking the time to write such a wonderful foreword.

Foreword

The book you hold in your hands is not inert matter, mere paper upon which ink has been spread in a particular font and format. The truth cuts deeper—a lot deeper—than what is visible to your physical eyes, tangible to your hands, and translatable to your mind. It's about what drew this book into your electromagnetic field in the first place: an energetic resonance between you and the wisdom-message upon its pages. This energy field is what Penney Peirce identifies as your "home frequency," or your personal energy vibration. The words on these pages vibrate at a specific energetic frequency too, sourced by the cosmic Intelligence enlivening and sustaining existence, and they can transmit to you their grace and power. When Penney describes how Spirit and Matter dance ecstatically together, you will be eager to learn the steps within her nine phases of personal transformation so that you may join in the celebration.

The world's scientific communities agree that energy comprises all things and that energy systems are conscious. Earth spins, they tell us, within an infinite electromagnetic field. Everything participates in this swirling, oscillating, vibrating energy. It's interesting how we commonly regard cosmic energy as something *out there*, crackling in some far-flung location beyond the earth upon which we walk. The truth is this same energy is present *in here*, right within our existence as an individual self and everywhere present in the atmosphere in which we live. We are spacious beings who live and move in such a way that energetically impacts every corner of the cosmos. So when we are dealing with the subject of energetic frequencies, it is not a mysterious "something" vibrating out in the stratosphere, but directly within our individual *inner space*. Pierre Teilhard de Chardin, the Jesuit priest and paleontologist, referred to this inner space as our "interiority," a word he coined as a result of his intuitive relationship with the natural world, a cosmology which he considered to be an energetic, constantly evolving progression toward material complexity and consciousness.

Chardin paid dearly for his theories: his work was banned by the Vatican and he was often unwelcome in his country of birth, France, so he moved to China and later to New York City. It is a testimony to our evolutionary progress that today individuals such as Penney are free to openly share the results of their inner explorations in the laboratory of consciousness without religious or governmental censorship and condemnation. Twenty-first-century breakthroughs in understanding the body-mind-spirit connection and our interconnectedness with the cosmos have shifted our views and opened us to realizing the Earth's innate intelligence. Humanity has evolved in its understanding and, as a result, is far more conscious about the energetic connection—indeed oneness—with our mother Earth and how we must live in harmony with the invisible, energetic laws that support all life. Penney has created what I'm choosing to call an "energetic template," which offers skillful means, practical applications to our everyday life, and the deeper aspects of what it means to be a spiritual being having a human incarnation.

Penney's detailed description of the Hindu *chakra* system of energies can give us a way to clear our energy field, to free coagulated energy. By applying ancient truths like these to the challenges of modern living, we can move from the Information Age to the Intuition Age, where the collective consciousness is becoming increasingly comfortable with our innate ability to tune in to the higher frequency of an "expanded self," what I call the Authentic Self.

Penney's brilliant scales of everyday vibrations go the distance in describing how we influence our body, emotions, and thoughts throughout the day. Together with her nine stages of transformation, we can apply path-cutting skills that attune us to the energetic frequency of our current vibration and then accelerate us to higher-octave frequencies. Penney is obviously committed to transmitting her knowledge with impeccability and generously shares with her readers a profound understanding of our energetic anatomy. She presents a winning case for the fact that we have not only the capacity but also the responsibility to calibrate and recalibrate our "home frequency" so that we may live our highest purpose.

How encouraging it is to realize that the application of intuitive energy is not reserved for psychics or mystics. It is a faculty we all possess and use, consciously or unconsciously, in varying degrees, depending upon our individual development. Penney gives us the good news that by practicing the exercises she teaches, we can consciously develop our intuitive faculty and thereby consciously draw energy from the original storehouse of insight, intuition, and inspiration—the very Life Source within and all around us.

Conscious cultivation and application of thought-energy is a most powerful agent for self-transformation. It caused Carl Jung to describe an experience he had at twelve years of age in this way: "Suddenly I had the overwhelming experience of having just emerged from a dense cloud. I knew all at once: Now I am myself! It was as if a wall of mist were at my back, and behind that wall there was not yet an 'I.' But at this moment I came upon myself. Previously I had existed, too, but everything had merely happened to me. Now I happened to myself."

When we awaken to our true nature as energetic beings, when we enter our interiority and begin a conscious exploration of the mystery of consciousness, we will "happen to ourselves." There are those who still consider self-reflection, contemplation, meditation, and other interior practices to be self-indulgent absorption. However, clinical studies and revolutionary advances within various disciplines of modern science continue to provide empirical evidence that inner practices positively shift the practitioner's energetic frequencies. Whether electromagnetic, gravitational, or quantum, science is revealing what spiritual giants of all times and traditions have told us since the beginning of time: we are luminous, energetic beings of creative intelligence, fully equipped to consciously participate in the evolutionary impulse of the universe and become fully self-realized.

In *Frequency*, Penney Peirce intelligently and compassionately combines powerful teachings, personal experiences, her work with clients, and skillful methods for upleveling our home frequency and uplifting not only our individual life, but life throughout the cosmos. Hers is a most excellent energetic medicine, universally applicable to the times in which we live.

Michael Bernard Beckwith,
author of Spiritual Liberation

To the Reader

The convergence of mysticism and the new physics has
brought us to the gateway of our humanness.
Beyond lies something that is literally
beyond our language.

Michael Talbot

You've no doubt noticed that we live in chaotic yet amazingly potent times. Just as restless animals sense an impending earthquake, you may feel a big change brewing. It's hard not to notice that everything today is as volatile as boiling water. The upside is that the volatility is shaking us up and pushing us to experience ourselves in an entirely new way—less as solid physical bodies separated by empty space and more as energetically vibrational beings living interdependently with other vibrational beings in a vibrational world.

We're becoming increasingly aware of internal and external *energy*, its qualities, and the principles by which it functions—frequency, vibration, resonance, waves, oscillations, cycles, octaves, and spectrums. We're discovering that these concepts are at the heart of the newest techniques for knowing, doing, and having everything. In other words, your *personal vibration*—the frequency of energy you hold moment by moment in your body, emotions, and mind—is the most important tool you have for creating and living your ideal life. If your energy frequency is high, fast, and clear, life unfolds effortlessly and in alignment with your destiny, while a lower, slower, more distorted frequency begets a life of snags and disappointments.

Frequencies Rising!

There are some key things to understand now: (1) you are being affected by an evolutionary process that moves through specific stages, which is causing the energetic frequency of your body, emotions, and mind to accelerate, (2) because the rising frequency of your energy parallels your level of awareness, you are gradually becoming more aware, sensitive, visionary, empathic, and loving, and (3) the biggest challenge of the next few years will be working with your sensitivity, keeping your personal vibration clear, and learning to use "frequency principles" to live success-fully in the coming times.

People who are sensitive to the invisible realms—and I count myself as one—have long intuited that the subtle frequency inside our bodies, and in the earth itself, has been steadily rising. This first stirs us up internally, causing us to feel uncomfortable without knowing why. Then the exter-nal world accelerates and seems increasingly high-pitched, or even chaotic. Eventually, we adjust to the new higher level of energy and our awareness increases to the same degree. We've known instinctively that the heightening energy was coming in a series of waves, carrying our awareness incrementally up toward a shift in perception, where our sense of self would evolve from an identity based on separation, fear, and ego to one based on interrelatedness, love, and soul. We sensed that at that high frequency, our world would function according to new, more elegant and efficient principles.

> Our style and manner of thinking have undergone a revolution. . .
> We see with other eyes; we hear with other ears; and think with
> other thoughts, than those we formerly used.
>
> Thomas Paine

Now this shift is under way and evident to most of society, as we try to function in a climate where everything is increasing: from the amount of data we must digest, to the hours we must stay awake to get the job done, to the tidal wave of negativity that's beginning to seem normal. It can be a daunting task these days just to stay centered! We're leaving the Information Age and entering the Intuition Age, which brings with it nothing short of a major transformation in the way we perceive reality.

The questions now are: How do we learn the rules of this expanded vibrational world and develop the energy and consciousness skills that can help us function in it? How do we stabilize our new perception, identity, and behavior while our previous way of life is going through its death throes?

Are the Frequencies Calling You?

Like many people, you may be responding to the accelerating frequency of life by trying to adjust your own energy state in a variety of ways—both healthy and unhealthy—to find equilibrium, security, and relief from stress. Or you may be hunting hungrily for clues about how to thrive in this excited world with its massive complexity. The answers do not lie in gadgets and gizmos or in technologically assisted ways of processing more data. The simple truth is that moving into the Intuition Age is all about what you can know and do with energy and how you can develop effective, expanded sensitivity.

You may have picked up this book because you'd like to stop being plagued by emotional tailspins that block the forward flow of your life. You might be drained by people who are disturbed and reactive or depressed and apathetic. Maybe you're overwhelmed by nonstop stimulation and don't want to continue to feel either numb or hyperelectrical. You may feel cluttered with subtle, nonverbal information you've picked up concerning other people, the news, the future, and the events in your life. You'd like to make sense of it but can't exactly pinpoint what's affecting you.

Perhaps you'd like to reclaim your sensitive, spiritual nature that's been lost amid academic, administrative, or materialistic pursuits. Your analytical mind may have brought you success in business, but now you may need to be wildly innovative, motivate people from within, and revolutionize systems that seem like dinosaurs. Have you made strides by understanding the "law of attraction," and do you want to know more about the new principles of our emerging reality? Are you trying to find the right balance between will and trust in creating your life?

If you feel nearly paralyzed by the challenge to be clear or change quickly, don't worry. Everything is proceeding in right timing, and we're all in this process together, all learning to adjust ourselves to higher frequencies of awareness being the norm. We're shifting from a world where we learned to use cleverness and willpower to bridge imagined gaps between ourselves and others—and get what we want—to a world where there are no bridges to cross, where love, support, the easy materialization of results, and freedom are readily at hand.

> The truth is that our finest moments are most likely to occur when we are feeling deeply uncomfortable, unhappy, or unfulfilled. For it is only in such moments, propelled by our discomfort, that we are likely to step out of our ruts and start searching for different ways or truer answers.
>
> M. Scott Peck

Your Highest Frequency Can Be Normal

I absolutely know that you can, under your own steam, dissolve the shell that separates you from a higher experience of Self and a much better life. You don't need gurus or to be catapulted into *super*natural experiences by dramatic events; you are becoming such a high-frequency being right here in your physical body that what used to be *meta*physical, *trans*personal, and *para*normal is now almost ordinary. Some missing pieces of the big picture are making their way into your consciousness now—and this new understanding is facilitating the "tipping point" into the Intuition Age for all of us.

Many people I talk to are close to understanding that we've never left Home—the "heaven" experience—while at the same time we've been having the most amazing, captivating Dream, called "life on earth." To fully wake up from the dream, you'll need an everyday experience of soul, of your own highest frequency state, that you accept as normal. That experience is, among other things, one of empathy and compassion, where high-quality feeling shoots you straight through and beyond the seduction of suffering, the limitations of logic, and the fuzzy hypnosis of the world. You have to *feel* lovable, loved, and loving—in your very cells—before you can grasp the truth of your enlightened identity, know oneness, and settle in to an expanded way of living. Being consciously sensitive to the subtle information encoded in energy frequencies will put you on the fast track to experiencing this saturated state of certainty about love and soul.

To discover the information and experience stored in our most refined vibrations, we are experiencing an evolutionary movement "down" from our heady view of life, into the knowledge of our bodies. Yet because we encounter emotion when doing this, and emotion can catalyze confusion, resistance, and panic, as well as lift us into a sublime mystical swoon, we tend to avoid it. By avoiding emotion we create a blank spot where we don't experience our full self. I wrote *Frequency* to help you melt through the last barriers to awakening fully and actually *feeling* the experience of the expanded self that will be your normal state in the Intuition Age.

There's More Beyond the Law of Attraction

In recent years, there has been a series of books and DVDs that have acted as a bread crumb trail to help us find our way through these stimulating times. They popularized leading-edge scientific and metaphysical ideas in a way that captured the heart and imagination of the public. Among these are *The Way of the Peaceful Warrior, The Celestine Prophecy, The Intuitive Way, Your Life Purpose, The Field, What the Bleep Do We Know!?, The Hidden Messages in Water, The Da Vinci Code, The Law of Attraction*, and *The Secret*. These works, and others, helped us clarify what we sensed was happening in the invisible realms and encouraged us to develop rudimentary skills in working with energy and awareness.

At the same time, because of the introductory nature of the material, the wide array of interconnected topics introduced in such a short period, and the climate of fear in the world, the content was misinterpreted in numerous ways. I liken this to laying out the dots and numbers for a paint-by-number drawing—once that's done, you still need to connect the dots before the skeleton of the artwork appears and you can apply your unique vision to filling in the drawing with your favorite crayons or watercolors. I wrote *Frequency* to help you complete the picture.

I Want to Help Ease Your Transformation Process

As a byproduct of my own intuitive practice, I've had many visions—starting in the mid-1970s—about what's happening behind the scenes in today's intensifying world. I recognized early on that we were experiencing the beginning of personal transformation the likes of which has not been experienced for thousands of years and perhaps never at a global level. In the 1980s, I began giving lectures about this process of heightening frequency with titles like "Predictions and Future Trends," "Holographic Perception and the New Paradigm," and "Eliminating the Gap Between Self and Soul." In these talks, I would sketch out the components of the transformation process, and afterward, audience members would tell me, "This helps—to have a broader understanding of what I'm experiencing. Without knowing the process has a higher purpose and positive outcome, I've been confused/scared/depressed."

A radical inner transformation and rise to a new level of consciousness might be the only real hope we have in the current global crisis brought on by the dominance of the Western mechanistic paradigm.

Stanislav Grof

So I want to help you understand this semi-invisible process you're being influenced by and help you move through the phases of it fluidly. I want to give you techniques to change blocked emotion into exquisite sensitivity so you can decipher the messages contained in the myriad vibrational states. I want to make it easier for you to hold your own, to choose to live from your highest, most natural personal vibration—the one that matches your own soul. I want you to know how to recenter into it when other people's lower vibrations drag you down. There *is* a way to become a healthy "sensitive," and there *is* great power in knowing how to work with your personal vibration.

Keeping a Journal Makes Your Growth More Conscious

Keeping a journal is a surefire way to track your growth process. You might document what happens as you penetrate into this body of information. Scattered throughout the chapters, you'll find a variety of simple exercises that can help you practice the concepts I'm giving you; try doing them and writing about your results. What insights did you have? What difficulties or surprises did you encounter?

You might play with *direct writing*, where you write straight from your core, letting a stream of words emerge as a spontaneous flow without censorship. Begin by posing a question, which serves as a magnet to draw forth a response from the deeper part of your awareness. Let the first words come; they will draw in the next ones. Don't think ahead or second-guess what's being said. If a strange word comes to mind, write it down. Whatever is supposed to follow will simply occur next. To keep the flow going, it's best not to read what you've written until it's finished. You'll be surprised what you find yourself writing because it will be so fresh and accurate.

> To know a thing, dip yourself in it like pen and ink,
> let it write you in its own words.
>
> Elizabeth Ayres

At the End of Each Chapter You Can Shift Your Awareness!

At the end of each chapter, I've included a piece of direct writing that came from my *home frequency*—my core self or soul vibration—while I was in a calm state of higher awareness. The voice in the messages is just the simple voice of our oneness—a voice we can all access equally; it's not some channeled entity. When I wrote these more inspirational pieces, my personality and normal writing voice were not in the forefront and the words came as the simplest, most direct description of deeper realities—or feeling states—that were being revealed *as I wrote*. There is little wasted breath in the words, and each sentence has the power to transport you. I've included these messages for several reasons:

1. They give you an example of the kind of wise, centering guidance your inner self can access when you become very quiet and sensitive. Hopefully, reading the messages will encourage you to experiment with tapping into your own clear source for direction and answers.

2. I want you to see that communication that comes from the heart and soul—anyone's heart and soul—as opposed to the cacophonous opinions of our ever-changing minds, has a universal appeal and contains universal truth.

3. The messages offer heartfelt insights that often extend the meaning of points made in each chapter. Like poetry does when compared to prose or technical writing, the messages take you beyond pure logic and let you see that not all knowing must fit neatly into little connected boxes to make sense.

4. I want you to experience the contrast between what it feels like to understand each chapter with your analytical left brain that likes to move fast and what it's like to intentionally shift to a slower pace so you can *feel into* deeper, less intellectual, more personal meanings with your intuitive "mind-in-the-heart." By experiencing this shift from surface to core, from the "older," linear-logical perception we're all so used to, to the "newer," all-at-once, experiential knowing that these pieces require, you'll begin to understand how you can enter the high-vibrational, instantaneous, transformed way of knowing the world that I'm describing in *Frequency*.

If you read these messages *much* more slowly and deliberately than you normally would, or read them aloud to yourself, or close your eyes and have someone else read them to you, you'll notice they convey a different rhythm and vibration than the regular text. They might sound corny—even New Agey—if you read them with a too-heady, skimming mind, but when you slow down so you can "be with" them and feel them in your body, they take on greater dimension and open you to new realities. Sometimes, holding one sentence or phrase in your mind a few moments longer than usual will reveal a hidden meaning. This is a mini-demonstration of the difference in the way reality can look and feel when perceived from the mind alone, and when perceived from the unity of mind, heart, and body.

You can skip over these *home frequency messages*, or you can use them to practice entering your own home frequency (which I describe in detail in chapter 5). You might want to read all the pieces in a row at the end; it really doesn't matter. What could be interesting is to see how you react to the change of pace, vibration, and focus the messages summon up in you. If you feel irritated—or relieved—when you come to one, for example, it might be a clue about how you're moving through the key transitions in your own process of personal evolution.

Using Your Intuition Helps You Experience More

It will help to pay attention to your intuition as you ingest and digest the material in *Frequency*. Intuition is direct, unbiased perception that comes from unifying the fragmented parts of your awareness—body, emotions, mind, and spirit—and it arises when you're focused in the present moment, alert but not tense, connected with your body and feelings, not using willpower, and feeling simple and open. Intuition can speak from different levels of your brain as instinctive attraction or repulsion, as impressions from one of your five senses, or as the sudden flash of understanding of a complex meaning, system, vision, or pattern of information. As I wrote in *The Intuitive Way*, "Our own private intuition is the catalyst for self-improvement and self-realization, because when it comes to making deep and lasting changes in one's personal life, it is only subjective experience, not facts, that registers as real."

When you pause between periods of reading and drift a bit, intuitive insights may pop to mind. When you're out in the world, experiences may occur that relate to the chapter you're reading. Intuitive "ahas" will make the information particularly real to you. Many times, you may need to shift into sensing or feeling how something works to fully understand the concepts presented. As you develop your sensitivity, it will be your intuition that brings you insights about the meanings in the vibrations you sense. Intuition is a doorway that reveals a higher reality and clearer experience of the Divine. And in the end, it will be your subjective interpretation of what your intuition brings that frees or inhibits your ability to act and grow.

Finding Frequency

The idea that the invisible universe is more real than the visible one indeed has never been so widely accepted by practical scientists as now in this climactic century. But it is far from a new notion to seers and philosophers, for don't forget, Aristotle called life "spirit pervading matter," a concept all great religions would heartily endorse . . . the philosophy of mysticism emerges as eminently reasonable . . . the newly realized reality of the nonmaterial world, of fields that influence, of waves that convey, of minds that pervade.

Guy Murchie

I've been deeply involved in intuition development and interested in what Buddhists call *skillful perception* (basically, how to use your mind to heal suffering) since the early 1970s. My enthusiasm for intuition and its related mysteries has never waned—and through my study of it, I've found similar unifying truths in all religions and spiritual paths and many secrets to maximizing an easy flow in life. It has been part of my intuitive practice to work with the "art of inquiry"—to regularly question what I know to see if it wants to dissolve or evolve into something more comprehensive, specific, or totally different. There have been times when an interest, such as reincarnation and knowing my past lives, which was a big part of my worldview for years, broadened into something so big and impersonal that it didn't hold the same meaning or importance anymore, and I stopped focusing on it so much. It's always surprising to be so fascinated with something, then so neutral—but intuition, I've learned, is most accurate when we remain honest and flexible.

In spite of my practice of inquiry, it hadn't occurred to me that there might be another stage beyond intuition development, a more specific skill that could take us deeper into our human potential. Recently, it dawned on me that I had not only become intuitive, but through my counseling practice, I had also developed empathic sensitivity to vibrations, textures of energy, consciousness frequencies, and patterns of interwoven emotion, belief, and purpose. And that working with all this was a next step.

We're All Evolving into Heightened Sensitivity

We are all born sensitive and empathic, but through lack of validation and training, the ability often shuts down or is placed on a shelf labeled "For Use at a Later Date." I'm lucky that mine stayed alive. I credit my mother for planting sensitivity seeds in me at an early age; we used to sit in the car waiting for my father while he ran into his office in Chicago on a Saturday, and to pass the time we'd play "What is that person thinking?" We'd read the minds of passersby and create imaginary lives for them. When disciplining me, she would often invoke the Golden Rule, saying, "How would you feel if you were in her shoes and someone did that to you?" I took the question literally and transported myself, via imagination, into my wronged friend's reality. I was learning to "feel into" people.

When I began working as a professional intuitive in my twenties, information came predominantly through my inner visual and auditory senses, and it was rapid and impersonal. Soon, my intuition shifted to a more tactile, sensation-oriented mode, what metaphysics terms "clairsentience," the inner sense of touch. I noticed I was sensing people with my whole body, not just my head, and I felt much more personally connected, though processing was slower. I began to feel the same way my clients were feeling. I'd look at a client whose face was stuck in a grimace on one side and feel my own face take on that exact musculature pattern. Within moments, I'd know the disgust he lived with. If I sat across from a client with a caved-in chest and rounded shoulders, I'd feel her inner posture in myself, which would elicit an understanding of why she felt sad or defeated.

> We are like islands in the sea,
> separate on the surface but connected in the deep.
>
> William James

While doing readings, I'd often feel the physical symptoms, say, of someone's angina or broken ankle, hear music if my client was musical, or smell flowers or the seashore if the client was oriented to scent. At times, I'd experience a texture of energy like sandpaper, or ashes, or silk and eventually realized that this was what the other person's reality felt like to them. I began to call this phenomenon "direct impress." I thought this was what being impressionable meant but at a very subtle level, as though information and feeling were actually pressing into me somehow. I also learned not to be handicapped by this way of knowing, since as soon as I focused on my own body and innate cheerfulness again, I would return to my natural, relatively well-balanced self.

> The whole outer world and its forms are a
> signature of the inner world.
>
> Jacob Böhme

Some years ago, a woman came for a reading who, unbeknownst to me, had suffered sexual and physical abuse in her childhood, and she was tense and surly. I had not experienced this enough to recognize it as the phenomenon it has now become, and as she sat militantly in the chair opposite me, arms crossed tightly over her chest, acting like she didn't want to be there, I felt intimidated and scared, then angry. I persisted with professional kindness, though, and started the reading. As I penetrated through her layers of defense, I discovered the abuse and felt how wounded and vulnerable she was. I saw that I'd been feeling *her* feelings a few minutes earlier; *she* was intimidated, scared, and angry. Then I felt her innate love under it all, and compassion for what she'd courageously experienced flooded through me.

I spoke to her about how I experienced her true self, and she broke down sobbing. I understood then how she had unconsciously mimicked the emotions of her abuser (a negative form of empathy or body sensitivity) and was acting out the same qualities he'd shown her but in a different way. I thought, "How many people have reacted to her as I almost did, reflecting back the lack of caring, rejection, anger, and hardness—and reinforcing her wound?" I learned a valuable lesson from her about how what I feel can be related to what people around me are feeling. We're just like tuning forks, copying the resonances we "touch" energetically.

I want to emphasize that this deepening from a primarily visual-mental orientation, to a more feeling-oriented mode, to an even deeper body-oriented ultrasensitivity was the developmental path that my abilities took, and it is not better than any other path. Some people remain more detached, working from the upper part of the brain, or receive information via one main, preferred sensory mode, like hearing. My path down through the levels of the brain and into a merger with the body has been my teacher, gradually taking me directly into the principles of personal vibration I'm so excited to share with you in this book.

You Can Know Many Things by "Feeling Into" Life

I've done tens of thousands of life and business readings. During the sessions, I relax into a soft, less-defined personal identity and expand to include more space and time, raising the frequency of my body, emotions, and mind to a higher level. As I stay focused on the person and match the frequency of my energy to theirs, I become aware of an equally expanded amount of them—a field of information that contains their potential. In effect, I merge with them, or *feel into* them. That energetic information then surfaces in me as though I am them, and all at once, I know a complex pattern that includes their earthly desires, soul intentions, permissions, blockages, life lessons, talents, and how it's likely to feel when they attain their potential. I know it through my own body sensations and feelings. The challenge then is to articulate the body of information in a logical, sequential way that makes sense to the other person. I have experienced many times how, at these deeper levels, it is impossible for us to truly be separate from others. So much is shared. And yet, paradoxically, we are also each unique. How all this works truly boggles the mind!

At times I feel as if I am spread out over the landscape and inside things and am myself living in every tree, in the splashing of the waves, in the clouds and animals that come and go, in the procession of seasons.

Carl Jung

The same process of *feeling into* can be used to deeply understand any-thing—a family dynamic, a sick houseplant, a corporation, a marketing trend, a country. Now, I know this is an ability we are all developing. Truly, it is our most normal way of knowing—this sensitivity born of the ability to commune with life. When we know and interact via expanded, higher-level vibration, the result is always understanding, appreciation, and compassion. I recount these things because you, too, may know about others in inexplicable ways or experience heightened sensitivity that floods you with information and insight you're not sure belongs to you. I want to reassure you that your sensitivity can become a strength and an incomparable skill for navigating the world, doing business, or attaining wisdom. In addition to deepening your knowing, working with the power of your personal vibration can help you clear your fears and be more loving, create true stability, have better relationships, manifest the life you need and want, and speed your spiritual growth.

How *Frequency* Gave Me Its Name

I love synchronicity because the experiences of alignment are reassuring and entertaining, and they draw my attention to a higher coordinating force operating in life. Though science tells us that coincidence is normal 20 percent of the time, I still pay attention whenever signs and omens occur. In the case of creating this book, I received a signal that penetrated right to my heart and demonstrated the principles I would be writing about.

My father lived alone in Florida, three thousand miles from me, and was full of conflicting emotions buried under a stoic façade. He never wanted too much display of affection or meaningful talk—I assume, for fear of losing control and turning into a blubbering idiot—so I always felt a bit of strain when connecting with him. In 2000, he died suddenly as he sat in his armchair after dinner: his heart gave out. A neighbor found him four days later. I couldn't believe I hadn't seen it coming, that I hadn't called or visited him beforehand. Perhaps he didn't want anyone to interfere with his peaceful passing and had blocked me from seeing his death. But I was still upset that I hadn't been there for him.

The day he died, I'd been writing a book on dreams and couldn't concentrate. I kept standing up suddenly from the computer and pacing around the house, then stopping and staring into space. What was I trying to do? Why was my creative flow blocked? By early afternoon, I decided to give up, play hooky, and go to the movies. I didn't know what was showing but arrived in time to catch a film called *Frequency*, starring Dennis Quaid and Jim Caviezel. In it, a son finds he can talk to his dead father over their shortwave radio due to strange atmospheric conditions. By communicating this way— by finding a common frequency that spans time and dimensions—they solve mysteries, heal old wounds, and create an improved future where the father doesn't die early. I figured I was watching that movie as my own father was dying, and that was the closest I could get to being with him in his passing. Ever since, that film has held an especially powerful charge for me.

Cut to the present. I'm outlining ideas for this book, and my mind is swimming with possibilities for cohesiveness, structure, and timeliness. What to call it? I'm filling pages of a notepad with a variety of clever phrases with modifying subtitles when my hand—by itself—writes FREQUENCY, all in caps. I stare at it. I feel my father very close, sense him grinning at me, like he's waiting for me to get the punch line of one of his numerous practical jokes. I think, "Yeah, that could work. Right, that could *really* work! But, no, it's too simple. But, no, it's a movie title. But, yes, it's really the right vibration." I shiver, sit with it, and realize I've just experienced the same common frequency spanning the dimensions between life and so-called "death" that the characters experienced in the movie. And my father is enjoying the layered meanings tremendously. I've just felt the power and magic of *frequency*.

> Don't ask yourself what the world needs; ask yourself what makes you come alive. And then go and do that. Because what the world needs is people who have come alive.
>
> Harold Whitman

Writing this book has been an education in itself for me. I thought I knew about vibration and sensitivity when I began, but the information and insights that spontaneously came to me in the process of writing were often mind-blowing. I've had so much fun penetrating into this body of knowledge and bringing it into words for you; in fact, the more I explored it, the happier I became. The visions and understanding I received about our coming reality are not in the least bit dire; they are shining and bright.

I welcome you to this fascinating investigation of the potentials that have been waiting patiently inside you for their time to come. This is that time, and I hope you enjoy it to the fullest.

Penney Peirce
Marin County, California

1

Our Phoenixlike Transformation

*There is a part of every living thing that wants to become itself—
the tadpole into the frog, the chrysalis into the butterfly,
the damaged human being into a whole one. That is spirituality.*

Ellen Bass

D o you ever feel like a balloon being inflated with helium? You're getting bigger and bigger, expanding to a huge fullness, and the gas just keeps coming? Try taking a minute to be aware of everything that interests, pleases, and upsets you. Then include all the creative breakthroughs, new facts, daily developments, bids for attention, and curiosities there are in the world. Now add all the opinions, complaints, hypocrisies, tragedies, dramas, and fears. Feel it all inside you, wanting to be known, pushing against the limits of the thinning membrane that holds your reality in existence. How much more room is there, how much more pressure can you take before your reality explodes? There's got to be another more unlimited way of knowing! Fortunately, we happen to be in the middle of an evolutionary process whose end result will be new kinds of perception, identity, and reality that are totally free of constraints.

There's a Good Reason You Feel Excited and Pressured

If you are to fully understand the emerging energy reality with its high frequency and unlimited perception, it helps to know there's a good reason you may feel shaken up, ultrasensitive, or rudderless. The reality of the Intuition Age will be the result of a gradual but fairly rapid process of personal and societal transformation. It can put you through the wringer emotionally and energetically but will eventually deliver you to an amazing destination you'll love. And, yes, there is a road map to help you get there. Understanding the transformation process as a whole is crucial if you don't want to jump from one popular trend in thinking to the latest energy technique while missing key components that can make your experience smoother, faster, and more cohesive.

In this chapter, I want to give you a summary of the transformational journey we have embarked on so you can have a framework for what's happening to you and the world. In subsequent chapters, we'll dive into an exploration of the power of your personal vibration to navigate the stages in the transformation process successfully, and you'll learn to develop healthy, conscious sensitivity that benefits both yourself and others.

> We are in a phase when one age is succeeding another,
> when everything is possible.
>
> Václav Havel

For years now, in the worlds of physics, business, spirituality, and even politics, we've heard about the *paradigm shift*, *quantum leap*, *new world order*, *tipping point*, *new age*, and *holographic reality*. Even *Star Wars* gave us a powerful image of the "jump to hyperspace." Thomas Kuhn popularized the concept of the paradigm shift in 1962, defining evolution as a "series of peaceful interludes punctuated by intellectually violent revolutions" within which "one conceptual worldview is replaced by another." So basically, a paradigm shift implies a change in the way we think about life that results in a change in behavior. Remember how developments in agriculture changed primitive hunter-gatherer societies, or the way the printing press freed us from church domination and the Dark Ages, or how personal computers and the Internet helped us shift from being isolated and local to being interconnected, global citizens? As significant as these developments were, they are within the realm of *change*. I believe what's happening today is more than change; we are living in a time of *transformation*. What's the difference?

Change + The Next Dimension = Transformation

Think of a dinner table with place settings, plates of food, beverages, and a centerpiece. Now mix up the arrangement—put the salt shaker inside an empty wineglass, the napkins under the plates, the food on the tablecloth instead of in the serving bowls, break a salad plate into pieces, and turn the centerpiece upside down. Voilà! You have change. You have recombined the existing items into new forms and relationships, but they all still exist on the tabletop in time and space, and there is still a perception that the items are separate from each other, solid, and inert.

Transformation, *metamorphosis*, or *transmutation* is something else altogether. These words imply an alchemical change in the basic nature of something, a shift from one energy state to another, a startling change that occurs miraculously, as if by magic. The phoenix legend exemplifies this "death" of the ordinary self by transmuting, purifying fire (spiritual growth) and its "rebirth" from the ashes as a creature of gold (light and love). In our example of the dinner table, the water might transform into its gaseous state and seem to disappear, or the whole dinner table might transmute into an experience of its higher meaning—abundance or an experience of family love—and cease to be a physical thing at all.

You're About to Experience a Big Identity Shift

Imagine a single point floating in space and you are that point. Life has no dimension, no movement, and you have virtually no sense of self. Now imagine a series of points gathering close together like beads on a string until a mysterious critical mass of awareness is reached and a new reality called "the line" is born. You merge completely into that new world, forgetting your identity as a point. You become a "line." You have an expanded self and live in a world with new rules: there is *movement* that oscillates back and forth. You are much freer than you were as a dot.

When my identity shifts, reality shifts simultaneously.
When reality shifts, my identity shifts simultaneously.

Now see if you can feel what occurs as a series of lines gathers close together and builds out in a previously unknown dimension, precipitating a new state of awareness called "the plane." You forget your identity as a line; you experience moving, not just in two directions, but anywhere in four directions—even in curves. Life as a "plane" offers you many more options, and you can barely remember your former limited self.

Next, try sensing what occurs when a series of planes gathers close together, stacks up, and expands into yet *another* dimension, catalyzing the reality of the three-dimensional "cube." You forget your identity as a plane because now you've opened into what seems like unlimited space and possibility. Your world contains time, space, and matter: finite objects and empty space; an inside and an outside world; a past, present, and future; and an idea of self based on what reflects back to you from other beings. Sound familiar? This "cubic," three-dimensional world is where we live now, who we think we are, and what we define as normal.

Here we are, somewhat discontented, highly excited, three-dimensional beings, closing in on the moment of the next evolutionary leap. What could transform your sense of self and world? It probably won't have much to do with computers that function in less *time* or traveling through *space* in three-dimensional spacecraft. The transformation of our world will become real when the next higher dimension, the fourth, integrates with our "cube" reality. You might think of the fourth dimension as the realm of the soul, of spirit—an experience rather than a place, in which everything exists simultaneously in a *unified field* of energy and awareness, where everything contains everything else, all is known, and "love" is the primary substance out of which everything is made.

What might reality be like when the third and fourth dimensions blend? Here's a hint: You won't be interested in linear cause-and-effect thinking or angular shapes. You'll think in spirals, and life will move in oscillating waves. You'll understand fractals and holograms as the basis of consciousness; there will be no such thing as distance, past, or future. In the new reality, soul will be the supreme force, consciousness of the interpenetration of the visible and invisible realms will be the most valuable thing, and anything will be possible.

Inca Prophecy Concerning the Coming Golden Age

In their book, *The Tenth Insight: An Experiential Guide*, James Redfield and Carol Adrienne quote Elizabeth Jenkins, director of the Wiraqocha Foundation: "The prophets of the Andes, the holy men and women, say the time period from 1993–2012 is a critical period in the evolution of human consciousness. We have entered the time they call the 'Age of Meeting Ourselves Again.' During these times, the Andean people believe we will make a transition from the third level of awareness to the fourth. The challenge is to cleanse our collective fears and gather enough spiritual energy so mankind can pass collectively into fourth-level consciousness."

Willaru Huayta, an Incan Spiritual Messenger from Peru, says: "The children of the Sun have existed since ancient times—since the last Golden Era. Even as there are four seasons of the year, so the four great cosmic ages follow one another. After the Golden Era came the Silver Era. Then the Bronze Age. Then the Iron Era, the present era consisting of the last thousand years. This last metallic era has a strong materialistic quality and has been an era of darkness, as people have fallen into egotism, using the forces of Mother Nature in a negative way. It is a time of wars, of the coldness of metal . . . The Age of Iron, like a long winter, is now closing. The new Golden Age, like spring, is announcing itself throughout the world.

"We must return to the Ways of Nature to receive illumination, to recognize the cosmic laws, and our bodies as temples. Each person is a Sacred Temple. The altar of that temple is the heart. The fire of love, a reflection of the greater light, burns upon this altar. This light within must be acknowledged, cared for, and venerated. This is the religion of the Sons of the Sun."

You Can Understand Transformation—
Here's the Miniseries

Over many years of working at a core, soul level with individuals and groups and seeking ways to understand the accelerated inner growth that people were experiencing, I evolved a highly detailed description of the transformation process. For simplicity's sake, I've distilled it here into nine phases. The process, however, is quite fluid and organic, with phases often blurring into each other. You may recognize how you've experienced all or part of each stage, or many or just a few of the symptoms I've detailed. You may have breezed through one phase and spent extra time in another, or bounced back and forth for a while between two or three of the phases. You may even feel you've completed the process only to find you've been affected by a new wave of even higher-frequency energy, and you're now moving through the process again to access more of your soul—but this time it's much easier. There's no fixed amount of time for traversing each phase; your process will be uniquely yours and the parts that are especially important to you will be highlighted. Eventually, everyone will experience the transformation process because it's something that's affecting the entire planet. The more conscious and proactive you are, the easier and faster you'll move through the experience.

It does not matter how slowly you go, so long as you do not stop.

Confucius

Finding the logic in the progression of stages can help you make sense of your life and current events. Note that each step naturally flows into the next, and at each stage you have a choice: you can either resist the change or trust that the flow is taking you into a better life. If you slow or stop the flow to try to maintain your old comfort level, you only stall your progress, reinforce fear, and create needless pressure and agitation. If you trust that a higher sanity is directing the flow and that it's just right for you and surrender to having the experiences you need, the frequency of your body, emotions, and mind will rise and you will experience the benefits of each phase.

Let's get a feel for the nine stages in the transformation process and look at the possible symptoms of the way each one may be showing up in your life and in society. The early phases of the process are concerned with clearing everything that blocks or interferes with your soul's expression and the experience of love. That means you typically become uncomfortable because you're looking at fear and what you've previously denied. These phases don't last forever, thank goodness! There's a turning point about halfway into the process where the light breaks through and you experience profound relief. After this, all disruptive symptoms disappear and you begin blooming like a flower, happily living your destiny. The transformation process begins when:

1. Spirit Merges with Body, Emotions, and Mind

The physical, three-dimensional world we think of as reality is beginning to respond to an influx of high-frequency energy, much as a napping child might be lured from sleep by the intoxicating smell of cookies baking in the oven. The fourth dimension, the next level of awareness that's just beyond you, is making itself known. Some say this is the result of distant cosmic events or a "plasma field" of cosmic energy passing through our part of the Milky Way. Whatever the cause, as the higher energy of this more spiritual dimension drops "down" in frequency and begins to permeate your world, you stir and respond, stretching "up" in frequency, yearning for the mystery.

I feel excited and know there's more to life! I want something better!

You experience thinning boundaries between heaven and earth—lightening you up, accelerating your perception and giving you the sensation of something important waiting behind a curtain. A process of interpenetration begins, coming in gradually intensifying waves, resulting in the eventual merger of the third and fourth dimensions. As you raise your consciousness and sensitivity levels, you catch glimpses of what's coming.

One of the symptoms of this first stage is the emphasis on mind-body unity that we see in psychology, integrated medicine, athletics, and spiritual practices, such as meditation and yoga. When mind and body become one, spirit or the soul is revealed, and you suddenly know it's been inside everything all along. You have a revelation about what it really feels like to be in your body *as your body* and how conscious every cell is, and you immediately know your body's instinctual response to any situation and why. You understand that everything physical is conscious in some way.

You may become interested in themes like attending church or spiritual groups, spiritualizing the mundane, being mindful, developing intuition, dreaming intentionally, finding your life purpose, and connecting with "supernatural" realms via crop circles, orbs, mediumship, religious mysticism, or near-death experiences. You might explore the subsurface foundations of life via physics, astronomy, microbiology, oceanography, and genetic research or be interested in energetic power places or buried secrets— things like ancient mysteries, new medicines, extinct species, missing-link life forms, or inner-earth "civilizations."

> Many people are now starting to experience a new energy filtering down
> through the density of mass consciousness. This energy stirs your spirit to
> find freedom of expression and amplifies the voice within your heart.
> This new planetary energy facilitates people in thinking more about
> the heart and its potentials in all human affairs.
>
> Doc Childre

2. The Frequency of Life Increases in Every Way, Everywhere

As the spiritual frequency penetrates the physical world, which includes both the planet and your own body, it also saturates your mind and emotions. Your body revs to adapt to the higher vibration, which initially disturbs your comfort level. High-frequency energy activates both positive and negative emotions, which makes you more aware of them.

Your heart may pound or skip beats. You may experience waves of internal heat. Not only is your body hotter and buzzier, but your feelings are more hyper and you may experience dramatic highs, lows, and erratic emotional releases. You can become overly sensitive and electrical and feel like you're under pressure that never lets up. You may see how your own "heating up" process parallels global warming. It's easy now to be intolerant of noise, crowds, allergens, certain foods, or the media.

**I feel shaky and nervous. Everything bothers me. My body aches.
I'm tired but can't sleep. I want what I want right NOW!**

You might experience a short attention span, short-term memory loss, lack of motivation, or disorientation. Physically, you can feel exhausted, be sick more than usual, or have more aches and pains. You might also feel hyperactive, impatient, irritable, and unable to relax. Insomnia is common, alternating with periods of sleeping like the dead. Many people retreat into their heads to the familiarity of logic and the desire for unlimited access to information. It's easy to become a slave to immediate gratification. You might try to escape your uncomfortable body by being absent-minded or living in a fantasy world, being distracted with problems and other people's lives, or obsessing and worrying. There can be an increase in higher frequency "overactivity" malfunctions of the body like cancer, viruses, fevers, infections, rashes, allergies, ADHD, and nerve disorders.

The positive affects of this phase are that if you allow the high-frequency energy to flow through you and let your body adjust naturally, you experience more vitality and endurance, as well as greater awareness, which means higher-frequency feelings (love, generosity, happiness, enthusiasm) and higher-frequency thoughts and motivations (innovation, creativity, inspiration, forgiveness, service, healing). You then yearn to know more, explore the mysteries, and experience the soul. You understand how positive thinking and love can heal, and you want to clean up your body (lose weight, detoxify, rejuvenate, exercise, move consciously and artfully).

3. The Personal-Collective Subconscious Mind Empties

As the vibratory rate of your emotions and physical body increases, subconscious blocks, which are fear-based emotions of low frequency, can no longer remain stored and suppressed. Low frequencies cannot exist in a field of high-frequency energy and awareness. The blockages respond like popcorn to heat, exploding up from their hiding places into your conscious mind. As they do, repressed memories surface and personal dramas and traumas occur, and old beliefs in limitation and negativity are unconsciously reenacted. You face shame, grief, terror, hatred, and the dark corners of your psyche. You must decide to undertake the hero's journey through the underworld, to penetrate into the subconscious and the unknown to find understanding. Naturally, there is great resistance to doing this, and courage is required. At this stage it helps to know there are proven methods for moving ahead, the phase doesn't last forever, and the things that are surfacing are not who you are.

Pessimism and dread tend to increase. Your dreams and fantasies can become intense, even violent and frightening, and your worst-case scenarios may happen. You experience a destabilization of things that used to be in balance, anxiety, and panic attacks, and you may think you're going crazy or you've developed a personality disorder. Scandals, taboos, abuse, and skeletons in the closet are revealed in public. Secrecy and privacy are things of the past. You might experience explosions—aneurisms, road rage, domestic abuse, acts of terror, and volcanic eruptions—as the lid comes off the pressure cooker. Or you might develop chronic pain as memories of painful abuse surface.

I'm upset and afraid. Life is too intense and won't let up.
My worst fears are taking shape. I can't control myself!

As your subconscious mind opens like Pandora's box, you notice increased dualistic, either-or thinking. Fear beliefs are based on polarity. For example, "If I'm loud, I'll be punished, so I must be silent." You now notice many polarities and their attached beliefs and emotions—good-bad, inside-outside, black-white, male-female, young-old, smart-stupid, beautiful-ugly, winner-loser, life-death. You learn how much willpower it takes to keep opposites separate, since the accelerated energy wants to connect them so they flow into, and turn into, each other—as in the figure-eight flow of the yin-yang symbol. When this happens, you experience the shadow side you've been resisting; then, as the energy flow moves you on into the light side again, you see how the dark and light, in their positive aspects, actually feed energy into each other. But initially, the pressure of the movement highlights the blocks to the figure-eight flow—your prejudices, fixations, and resistances—wherever they exist in you.

Of course, it's easier to see your unconscious fixations as they appear in other people and to think they *only* exist in the other: "I have to be quiet, so you're bad for being so loud and needing so much attention." You can't tolerate, and want to reject things and people you feel are different from you—in looks, character, intelligence, skill, or energy level. You may feel threatened, betrayed, or jealous. You may blame and judge yourself and others, and become caught in endless flip-flopping in better-worse, attractive-repulsive, aggressive-defensive, or possessing-rejecting polarities.

There is an increase in argumentativeness, name calling, us-versus-them scenarios, and condemnation. Relationships, especially those based on lack of honesty and communication, have trouble, reach impasses, and break apart due to blame, criticalness, or refusal to change. You may experience divorce, legal conflict, problems with neighbors, family disputes, or a craving for the perfect soul mate. These things become more visible: hate crimes, torture, voyeurism, indulgence in taboos, reality TV, forensics and crime detection, partisan political pundits, caustic talk show hosts, and psychotherapy. There can be an increase in chaos- and panic-related ailments like asthma, schizophrenia, bipolar and borderline disorders, or epilepsy.

The positive results of this phase are that if you allow the higher energy to flow through you and deal patiently and lovingly with situations that arise, you learn that you can't avoid polarities you don't like, but must allow every option to be part of life. You embrace the concept of "mirroring," where what's in you is also in me in some way, and vice versa. You learn to own both sides of any polarity, feel how they feed each other and how you can receive energy and information from parts of yourself you've rejected. You gain awareness about your previously unconscious emotional "triggers."

Even if you live to be 100, it's really a very short time. So why not spend it undergoing this process of evolution, of opening your mind and heart, connecting with your true nature—rather than getting better and better at fixing, grasping, freezing, closing down?
Pema Chödrön

4. You Retrench, Refortify, Resist, and Resuppress
Just as it looks like you might relieve yourself of the fear you've carried for years and heal emotional wounds once and for all, your ego—the aspect of

your mind that's rooted in fear and self-protection—steps in and screams, "I'm not ready to die!" It fights for preservation of the old difficult-but-familiar ways and vehemently avoids letting go and trusting. You plunge into survival behaviors—either fighting or fleeing—the ingenuity of which is staggering.

I do not have to change!
I am in my own world and I'm right!

If you're a fighter, you might make yourself feel invulnerable and large in a tanklike SUV or opulent home, or take on the world with aggressive workaholism and fixed belief systems, or drink lots of strong coffee. You might champion your personality, rugged individualism, and patriotism, and be more self-righteous, boastful, narcissistic, enraged, or even violent. You might want to be famous, powerful, or rich, or have cosmetic surgery and flashy possessions. You might seek security by controlling your environment and others, and when you realize you can't control everything, you might assume an attitude of toughness, cynicism, sarcasm, or belligerent apathy and disrespect, making it cool to do inappropriate, unkind, and amoral things if you want to. Patriarchal power structures like government, the military, business, and religion desperately and cleverly exert influence over others and maintain control with seduction, hypnosis, bald-faced lies, and magnification of fear.

People are driving me crazy. They should just go away!
I feel trapped in conflict.

If you're conflict-avoidant, you may resuppress threatening ideas, resist change, crave safety, or retreat into a perfectly decorated cozy home with a giant television or into relationships with parents or authoritative parental substitutes. You might eat more or less, gain weight with comfort foods or become anorexic, and numb yourself with audio and video piped directly into your head via tiny gadgets. Maybe you pretend there's no problem by maintaining a cheery persona, or escape into nostalgia for the past, future trends, fantasies of other worlds, or even become suicidal. You may feel you're finished with life on earth and want to join the angels or extraterrestrials. There's an increase in underactivity and resistance ailments like obsessive-compulsive disorder, stroke, bone disorders, arthritis, organ failures, addiction, depression, various kinds of paralysis, and exhaustion syndromes.

The positive results of this phase are that you're likely to experience sudden breakthroughs in all areas of your life, including healing insights about your wounds. Problems stemming from a difficult childhood can actually evaporate, isolation can give way to nurturing cooperation with others, and things that used to make you react vehemently now hardly bother you. You see through the chaos that others are caught in, don't buy in to the seductions, and find your own way forward with greater clarity.

5. Old Structures Break Down and Dissolve
We can only resist the inevitable for so long. As we experience ego death at a personal and societal level, many people panic, thinking this is the end of the world. Yet it is not—it's just the snake shedding its skin. What's really happening is that the way you identify yourself is changing from a limited sense of self to a very expanded one. If you act in an isolating, self-protective, controlling, or attacking way, these methods create instant negative repercussions now. It's good at this point to seek help from a friend, therapist, minister, guru, twelve-step program, or the angels.

I need to change because nothing is working.
I have to let go. I don't know who I am!

As you learn to clear your fear-based past, many things you thought were important and ways you made life meaningful become useless, even boring, and you let them go. Relationships based on codependence end. Old methods fail to produce results. Old habits die and old institutions become hamstrung and collapse. You notice more lies, disbelief, hollow stories, boring opportunities, untalented people, artless art, and pathetic attempts to cover personal weaknesses. You may feel disgusted. Your pet ideas, beliefs, and worldview won't hold water. You're sick of hearing yourself recite the litany of your own history and feel limited by it.

If you cling to people, possessions, situations, ideas, or habits, you are forced to let go. You may experience dramatic financial losses or bankruptcy, or lose your job, house, friends, dog or cat, or family. You may have more deaths in your life than ever before. Your well-laid plans might change because of circumstances beyond your control. Similarly, any ways society acts egotistically—dominating politicians, outrageously paid CEOs, the worship of privileged celebrities, or the church protecting pedophile priests—must fail and give way to new high-frequency behaviors.

> When the higher flows into the lower, it transforms
> the nature of the lower into that of the higher.
>
> Meister Eckhart

As old forms dissolve, you might experience disillusionment. You may have little certainty left about who you are, what you can rely on, or why you're here. You might have fuzzy boundaries and be open to invasion—by germs, parasites, allergens, dominating friends, thieves, terrorists, and non-physical entities. You may be forced to stop, perhaps by falling or having an accident or injury. Your body can be susceptible to loss-of-control ailments, such as fainting, diarrhea, Parkinson's, Alzheimer's, or multiple sclerosis. Other themes to surface might be: stock market crashes, public figures who fall from favor, chaos theory, black holes, death and life-after-death, Armageddon, reincarnation, transformation, the hero's journey, monsters and otherworldly creatures, ghosts, angels, forgiveness, fasting, and spiritual healing.

The positive results of this phase are that if you allow the dissolution of the things you don't need, you find that external rules aren't so necessary, that you're being directed from inside yourself and you see that you're living by inherent universal principles of natural harmony and order. You experience being guided effortlessly by your own highest wisdom. The failure of old structures is a natural way that life is preparing you for your new self.

6. You Stop, Let Go, and Relax into Your Truest Self
You finally reach a point where you stop fighting and struggling. Nothing's working. Perhaps you hit bottom or a mysterious enlightening

moment sweeps through you one day, revealing the simple truth. You are sucked entirely into the present moment away from your controlling ego. You can't make yourself do the things that used to work so well, and you experience the first real sense of your soul's vibration. There are simplicity, spaciousness, quiet, freedom, and peace. As you first encounter this state, however, it may feel like emptiness and your mind may panic and jump away, back to busyness and comfortable behaviors and ideas. As you let things be, you experience the profundity of your own being, and the experience transforms into one of relief, grace, and ultimately, joy. Suddenly you know who you are with your whole body! You've let go, found true center, and it feels great! Ego? Who needs it! You're fine, fantastic, in fact—just as you are. You've reached "the end of progress." Doing more is not the answer.

Now you enter a period during which you may have very little motivation, feel like you're in limbo, crave time in nature or by yourself, or question everything you've done so far and your goals, as they no longer seem to fit or be interesting. You don't fall into victim thinking about these things; you're much more neutral, as though you're a scientist observing some alien life form. You focus on unconditionally accepting yourself, others, and life, and on letting go and trusting—everything. This is a time of ripening in which you saturate with the frequency of your soul and receive clear yet subtle signals directly from your core. It helps now to enter your body in a fuller way, to activate sensory and artistic awareness, appreciate beauty and simple pleasures, and take action that allows innocent childlike involvement.

When I let go, it feels good! I can feel my real self everywhere and I love it! In fact, I prefer it, and from now on I refuse to sacrifice this experience for anything.

In the relative stillness, you may feel that your priorities, belief systems, and molecules are being rearranged, that you're being "rewired." You recognize the last remnants of what's untrue about your life and determine to disengage from participating inauthentically in the world. It may be obvious that you don't fit well with the rest of society and you must resist the pressure from others to revert to familiar habits.

It is key to find a "felt sense" of your soul now—an experience of your *home frequency*, or your highest personal vibration—so you can repeatedly choose it and recenter into it whenever you drift too far and become confused. We'll explore what this is and how to do it in chapter 5. As this experience of being "at-home-in-the-center" becomes routine, you entitle yourself to prefer your new self and new reality; this is the crucial turning point at which you intentionally choose who you want to be and what kind of world you want to live in. You become interested in things like the power of now, the law of attraction, meditation, authenticity, the soul, prayer, blessings, renewal, vision quests, and all forms of spiritual practice.

The positive results of this phase are that once you make this choice, the tide turns and your life, health, and happiness improve dramatically. You see how much better you feel physically and emotionally, how much easier it is to be creative and successful. You're receptive to insight, remember long-forgotten truths about yourself, and gain great understanding, often all at once. Now it's just a matter of practicing the new habit of staying in your home frequency and managing the level of your energy and awareness.

> The way of the Creative works through change and transformation, so that each thing receives its true nature and destiny and comes into permanent accord with the Great Harmony: this is what furthers and what perseveres.
>
> Alexander Pope

7. You Re-Emerge into the World like the Phoenix

After the big turning point in phase 6, there are no more disruptive effects on your energy and awareness. It's all "up" from here! Part of identifying with your soul's vibration is that your perception shifts from old self to new, and life behaves differently. Now you don't need to use willpower to make life work—you see that it's already working perfectly. You understand that you're interconnected with everyone and everything in a mutually supportive state and that life helps you do and have whatever you need. You become interested in self-fulfillment and self-responsibility, are "full of yourself" in the best sense, and want to do something innovative that comes from the cocreation of your body, mind, and soul.

I remember what I'm here for and now I want to get on with it!
I'm excited about my own destiny!
I can easily hold my new frequency,
even when other people vibrate at a lower level.

As you see others who haven't yet made the shift, you don't fear them as powerful wet blankets who can bring you down. Instead, you take the role of teacher, healer, or mentor and use your higher frequency for the greater good. You are hopeful, enthusiastic, optimistic, faithful, and inspired. Innovation abounds. You're motivated to recognize your destiny, engage with it fully, and live it. You look deeply for the work and self-expression you're "built for," honor the deep tendencies that have been present in you from birth, and eliminate any driving motives so you can find natural, effortless expression. Now it's a top priority to develop intuition, keep your heart open and soft, and discover surprising new directions, locations, and benefactors that occur in the absence of fear. You receive support, messages, opportunities, and miracles, and feel deserving and encouraged.

Your most-real motive is not to accomplish goals but to be multifaceted and so in the flow that you might be able to *shape-shift*—in other words, change into something radically different—if life wants to send you in a new direction. Materializing goals is easier and more fun, and your concept of time changes completely as past and future disappear into an expanded present moment.

Transformation literally means going beyond your form.

Wayne Dyer

8. Relationship, Family, and Group Experience Are Revolutionized
Now you *feel* how intimately individual and collective consciousness interpenetrate each other and know yourself as both a personal self (me) and a collective self (us). You feel how you and others affect each others' lives. This makes you want to be responsible for your actions and thoughts as a way to be kind to others. You don't have to remember to practice the Golden Rule because it's painful not to do so.

I can use my new frequency knowledge to
expand my relationships exponentially!
Other people love to help me, and I love to help them!

You're more tolerant and humanitarian. You see similarities between people and can hold differences as interesting and valuable. You work cooperatively from a feeling of fellowship and kinship, see your relationships as aspects of yourself, and are able to freely give and receive as the flow dictates. This has a profound effect on your sense of abundance and what is provided for you, empowering you to be more imaginative, creative, and productive because you know you have help and you know others need your help.

You develop a new understanding of yin/receptive and yang/dynamic energy, as well as the right and left brain. You balance yourself with equal development of your receptive-intuitive-sourcing side and your active-focusing-creating side so your perception can be more fluid, creative, and perpetually renewed. As you understand the balance of yin and yang awareness and energy, you draw parallels between these qualities and male-female gender dynamics in relationships. You find a new awareness about what's possible in relationships between two internally balanced people, along with new behavior possibilities for both genders.

Mayan greeting: "In La'kech" (pron: ein lah kesh).
It literally translates as "I am another yourself."

You are more calm and compassionate concerning the forming and ending of relationships as you feel the souls' purposes for coming together and moving apart. You may often be overwhelmed by the sheer magnificence of love. You may notice a broadening of the definitions of marriage, family, teamwork, organizations, and even international politics. Collaboration, cross-pollination, sharing, role reversals, and a shift in profit motive occur, along with a proliferation of new partnerings, networks, and rapid globalization.

You are authentically yourself, without having to struggle, and at the same time, you can work with the merged *group mind* in any team to arrive at more complex, complete, high-frequency answers, innovations, and joyful social experiences. You learn to modulate your personal frequency to match other people, groups, places, time periods, and dimensions of awareness, increasing your understanding and wisdom dramatically. This enables you to stretch into actual contact with nonphysical beings, spiritual councils, your soul group, and people who have died. You become adept at what previously were considered paranormal or supernatural consciousness skills that involve bridging and resonance, such as telepathy, teleportation, clairvoyance, spiritual healing, and psychokinesis (moving objects with your mind).

9. Enlightenment Is Grounded in Every Bit of Matter

With inspired collaboration and cocreation come shared wisdom and unbounded appreciation. Love is truly understood as *the* force of nature consistent throughout the cosmos, capable of miraculous things. When helping others is what you choose to do with your freedom, great nurturing is experienced by all.

> I feel free and unlimited! I can come and go through
> time, space, and beyond as I please.
> I can create anything and know anything instantly!

Personally, you take positive steps to act on your destiny impulses and find childlike joy in achieving results based on the soul's wisdom. When you're in the moment, in the heart, in the body, and connected to all knowledge, energy, resources, and collaborators, you understand the power of the unified field to help you materialize your visions. You discover how the creation-materialization process works and experience the unified field as an extension of your own body. Since the outside world isn't separate from you, your visions, goals, resources, and results—what used to be in the future or in another location—are in the moment with you, and thus, they can materialize and dematerialize in response to thought. Life isn't just fast; it's instantaneous, yet you don't feel pressured. You know how to work calmly with the natural filtering system of the present moment and the lens of your conscious mind and body.

You learn that your destiny evolves based on your interconnection to other souls and their destinies, and you discover new ways to plan and achieve goals in a fluid world. Birth and death lose their meaning as the great punctuation marks in life; you have the experience of coming and going as you please via ascension and descension. Your body is much lighter and more transparent and you know the experience of heaven on earth.

Try This!
Where Are You in This Process?

Review each stage of the transformation process and make notes in your journal about:

- What symptoms of each stage have you experienced?
- How did you contract or resist the process at each stage? What sort of repercussions did you experience?
- How did you surrender and cooperate with the flow at each stage? What benefits did you experience?
- What is your leading edge right now in the process?
- How do you sense you might be blocking yourself from moving to the next phase of growth?
- Where are other people you interact closely with in their transformation process? By knowing this, how can you understand them better and be more compassionate toward them?
- How do you notice the various phases of the transformation process acting out in current events or on the world stage?

> Personal transformation can and does have global effects. As we go,
> so goes the world, for the world is us. The revolution that will
> save the world is ultimately a personal one.
>
> Marianne Williamson

Just to Recap . . .

You're starting to see life as energy and awareness. This causes you to be ultrasensitive and aware of vibration because you are in a process of becoming a high-frequency person, of transforming from a relatively dense body and personality into nothing less than your soul fully saturated into time, space, and matter. You, along with everyone and everything else on this planet, are evolving through nine phases of growth that are designed to effectively raise your vibration into the fourth dimension— where spirit and matter merge—without having to die to do it. At first, this process may cause you to feel disoriented, frightened, or uncomfortable, but if you work consciously with honing your sensitivity and choosing your soul's vibration, you'll float through the transformation process like a leaf carried downstream by a river that knows where it's going. The phases are:

1. Spirit Merges with Body, Emotions, and Mind
2. The Frequency of Life Increases in Every Way, Everywhere
3. The Personal-Collective Subconscious Mind Empties
4. You Retrench, Refortify, Resist, and Resuppress
5. Old Structures Break Down and Dissolve
6. You Stop, Let Go, and Relax into Your Truest Self
7. You Re-Emerge into the World like the Phoenix
8. Relationship, Family, and Group Experience Are Revolutionized
9. Enlightenment Is Grounded in Every Bit of Matter

Home Frequency Message

As I explain on page xxi in *To the Reader*, I've included these pieces of inspired writing at the end of each chapter as a way for you to shift from your normal, speedier reading mind to a deeper kind of direct experience. Through these messages, you can intentionally change your personal vibration.

The following message is meant to transport you into a way of knowing the world that's close to the way you'll experience life in the Intuition Age. To move into the *home frequency message*, just downshift to a slower, less hurried pace. Take a slow breath in, then out, and be as calm and still as possible. Let your mind be soft and receptive. Open your intuition and prepare to *feel into* the language. See if you can experience the deeper realities and feeling states that come alive *as you read*.

Your experience may take on greater dimension in direct proportion to the amount of attention you invest in the phrases. Focus on the words a few at a time, pause at the punctuation marks, and "be with" the intelligence delivering the message—live and right now—to you. You might speak the words aloud, or close your eyes and have someone else read them to you and see what effect they have on you.

START BY BEING PRESENT

Just be: right here, right now. Listen for the quiet. Feel the relief. There is nowhere else, nowhere to go. You are surrounded by openness, and in that space is: awareness. It seeps through you—it is your own refined presence, your next level of self, the presence of the Divine. The awareness contains everything you've ever known, been, and will be, and everything everyone else has been, is, and will be. You are centered in the open heart of Love, in a vast field of Truth. Here you are real, here you are continually being born. There's nothing you have to do. Feel how the all-knowing, all-loving, all-supporting awareness: has you. It will never abandon you. You are safe.

Here you have unlimited energy and imagination. If thoughts occur, you don't own them—you're just becoming aware of things floating around, and if they seem interesting, you stop them for a while. You can identify with them, or let them go, or you can add energy into them and shape them and then let them go. There's no right thing to do, it's just creativity, it's just fun, it's just your soul expressing itself. So be in the moment, be soft. Let realities come and go. Your energy and awareness radiate through and beyond your skin, out and out, and you don't end, you just discover different types and patterns of knowing. As you include these in yourself and blend with them, you experience yourself in new ways. You are in all things and all things are in you, and whatever wants to be known or created simply appears in you as an idea, or expresses through you as an act. You can't make it happen: it's happening.

When you leave this experience and the moment, you will feel separate from the rest of life. And you will be sad, because you will miss yourself. You will miss the experience of your own soul expressing through your body, enlightening, enlivening your body and making it happy, just happy to be. You will not see the self you love in others either, and you will suffer. Meanwhile, in the center of everything, in the center of the awareness that runs through the world, there is always: what you're looking for. There is your answer: to everything, freely given, waiting for you to fall back into it. In this moment: the answer, the just-right answer for you, right now.

2

Living Among the Frequencies

The universe is more like music than like matter.

Donald Hatch Andrews

Do you ever wonder, while you're driving or sitting at a coffee shop, how many invisible waves and vibrations are crisscrossing through space around you? Are you aware of the radio programs and telephone conversations whizzing by your ears undetected? Or of the ambient fields of electromagnetic "pollution" you've encountered today? What about the physical pains and emotional moods of the people near you? There's so much invisible movement and information being broadcast and transmitted! I stood up from the computer the other day after hours of writing, and while walking to the kitchen, I felt the exact moment I left the computer's electromagnetic field, which extended a good eight feet. I'd never noticed this before. The vibration in the kitchen was calmer, yet when I stepped out the back door into my garden, the vibration was even smoother and more soothing. I realized how, after spending most of the day every day in the computer's force field, I had developed a kind of coherence with that subtle, even prickly vibration—so different from my body's more refined tone. And it was obvious at once why I'd been having so much trouble falling asleep the past month or so.

Sensing Waves and Sorting Vibrations
Is the Next Big Skill Set

As you become more sensitive, you, too, may be starting to feel the fields of electricity and magnetism, or discriminate new disturbing or soothing energies, or notice that the stranger next to you is about to get married or is developing an illness. What's more, you may be receiving this information *directly from the frequencies*: no radio required. This heightened, direct sensitivity is a fairly new ability that's surfacing these days—with no prior preparation or training. It can be disorienting when you realize you're beginning to know things you didn't know you *could* know, and you don't really know *how* you're doing it.

In the near future, you're going to be able to sense many more subtle influences in your life. You'll feel the fields of vibration you enter and leave, know what's healthy and unhealthy, and know when other people are on their game or off. You'll feel when an "event wave" begins to impact you before the event happens, and when a wave of energy is changing to a new frequency and you need to adapt to stay attuned. Your body will tell you when something isn't going to work, when guidance is knocking on the door to come in, or when problems arise at a distance and others need help. With heightened sensitivity, you'll be able to sort the vibrations coming from many sources to know which are true and actionable, and radiate intentional vibrations to achieve specific goals.

You may already do many of these things without consciously realizing it, but working intentionally, in a detailed way, with the principles of waves, cycles, spectrums, and fields is going to be a big part of your new skill set. Part of this will be developing your capacity to sense what energy is doing and being fluid enough to adapt to changing flows and rhythms without losing your center. It's always important to first learn the variables of the game you're playing, then master your game skills, and finally set yourself loose to play your best game. In this case, because we're aiming to become energy and consciousness experts, we need to be familiar with the varieties of frequencies outside and inside us, as these are our game variables.

> The world is never quiet, even its silence eternally resounds with the same notes, in vibrations that escape our ears.
>
> Albert Camus

When I told a colleague who works with leaders in government, business, and innovation that I was writing a book about frequency, he said, "I hope you're not going to throw the term 'vibes' around like some kind of New Age jargon. No one seems to be grounding ideas in physical reality these days." I took that to heart, so in this chapter I want to summarize a few of the important "real" vibrations and spectrums of energy affecting you—not because you need to become a physicist or an electrical engineer, but because I want you to feel and imagine yourself as a porous, vibrational being merged into a vast field of vibrations; we are not solid lumps on a rock of a planet, but a collection of energies penetrating and being penetrated by millions of other energies. I also want you to understand that physical energies, like the electromagnetic spectrum or sound and heat, have parallels in a higher spectrum of consciousness that contains different frequencies of *awareness*. In subsequent chapters, we'll learn to work with the principles of waves and fields to become more consciously sensitive, but now let's turn our attention to the sea of frequencies we live in. As I make a rough collage of some of the frequency domains, you might contemplate how they fit together and overlap, how you're affected by them, how you might become more sensitive to them, or how you might develop and use them to create a fuller life.

Outside You, the World Is Vibrating

Albert Einstein gave us a great truth: $E = mc^2$. I distinctly remember the moment I read this in physical science class as a kid—the idea that mass or matter could really be very slow, compacted, stored energy, that matter and energy were versions of each other. It made my mind go ding-ding-ding, like a pinball scoring points. A rock is energy, a cookie is energy, a log in the fireplace is energy, my body is energy, I am energy! Everything is moving at different speeds, and matter might convert into other more or less active forms of energy—water to steam or ice, for instance. So what might I convert into?

Then we learned that down inside seemingly solid objects are worlds of molecules and atomic particles, vibrating, rotating, and orbiting. And inside those particles are smaller subatomic particles. Now quantum mechanics reveals that those tiny particles of matter are also waves of energy, that both matter and energy can act as either a particle or a wave. In other words, the two states, as Einstein indicated, actually *do* become each other. What's more, if you look for a particle, the *wavicle* or *quantum entity* (e.g., a photon, electron, or neutron) becomes a particle, and if you look for a wave, it becomes a wave. First, this points to the fact that you are not separate from the world that you observe and define as outside you. Your perception determines the shape of your reality.

An atom is not a thing.

Werner Heisenberg

Second, energy and matter don't exist together in reality, only in probability. If we measure a particle's position we can't know its momentum, and if we find its wave movement, the position blurs. Thus, quantum entities are things that might be or might happen rather than things that are. The result is that a quantum entity exists in multiple possible realities called *superpositions*. As soon as an observation or measurement is made, the superposition becomes an actual reality, or the wave function "collapses." The many become one. Any given moment contains unlimited futures that can become real. The reality that occurs is the one you pay attention to.

Now here's the kicker: according to the *many-worlds theory* in physics, our world is split at the quantum level into an unlimited number of real worlds, unknown to each other, where a wave, instead of collapsing or condensing into a specific form, evolves, embracing all possibilities within it. All realities and outcomes exist simultaneously but do not interfere with each other. This gives some basis to metaphysics' concept of *past and paral-*

lel lives, that for any path of action you choose, there are dozens of other yous living different versions of your soul. All this makes you wonder: why do we think we're so solid and finite, or that miracles are impossible, or that radical change is foreign to our basic nature?

The new formula in physics describes humans as paradoxical beings who have two complementary aspects: They can show properties of Newtonian objects and also infinite fields of consciousness.

Stanislav Grof

We Live in a World of Waves

From matter's core wavicles, to the sunrise and sunset, to the invisible frequencies that carry information, the world around us is oscillating—rocking in and out, back and forth. Perhaps when we look at life as a conglomeration of particles, it freezes into solid matter and fact, and when we perceive it as a series of waves, it melts into an ocean of energy and potential. Now we see it, now we don't, and—surprise—now it's back again! Certainly, science has grounded us in the reality that our external physical world is vibrational (in other words, composed of a wide spectrum of energies—some perceptible to us, but most imperceptible). We know that energy moves in waves, that waves come in a range of amplitudes (intensities) and frequencies (speeds), which gives them unique qualities and behaviors. We also know that energy waves travel through a medium, or *field*, like air, water, or even awareness.

Our electromagnetic spectrum, the measurement of one of physics' four basic forces (electromagnetic, weak, strong, and gravitational), contains waves as long as 100,000 kilometers down to those that are a fraction of the size of an atom. Our language can barely describe it; we make a feeble attempt, much as we might describe the sizes of our café lattés, by labeling the frequency bands of just the radio wave spectrum as extremely low, super low, ultra low, very low, low, medium, high, very high, ultra high, super high, and extremely high. Actually, electromagnetic radiation can be divided into octaves—as sound waves are—resulting in a total of eighty-one octaves.

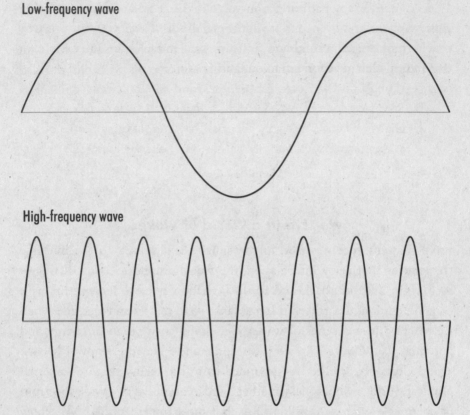

Low-frequency electromagnetic waves have a long wavelength and low energy, while high-frequency waves have a short wavelength and high energy.

The low-frequency end of the spectrum builds from radio waves, to microwaves, to terahertz radiation (T-rays), to infrared radiation, to the visible light spectrum, to ultraviolet radiation, to X-rays, and finally, to gamma rays. In case you aren't too familiar with the function of all these vibrations, here are a few interesting tidbits.

Radio waves are used to transmit data via television, radio, shortwave radio, mobile phones, MRI, and wireless networking.

Microwaves cause certain molecules in liquid to absorb energy and heat up, as in microwave ovens, while low-intensity microwaves are used in Wi-Fi.

Terahertz radiation is used in imaging and communications, especially in the military, as it has the ability to penetrate a wide variety of nonconducting materials.

Infrared radiation is used in astronomy and in night vision/thermal imaging technology, since hot objects radiate strongly in this range.

The visible light (rainbow) spectrum is the range in which the sun and stars emit most of their radiation, and it corresponds with our world of sight—though birds, insects, fish, reptiles, and some mammals, like bats, can see ultraviolet light.

Ultraviolet radiation is highly energetic and can break chemical bonds, making molecules unusually reactive or ionizing them. Sunburn, for example, is caused by the disruptive effects of UV radiation on skin cells.

X-rays pass through most substances, which makes them useful in medicine and industry, as in radiography and crystallography. They are also given off by stars and nebulae and thus used in high-energy physics and astronomy.

Gamma rays are produced by subatomic particle interactions and are actually high-energy photons. They have great penetrative ability and can cause serious damage when absorbed by living cells.

Here's what the whole electromagnetic spectrum looks like:

In case you were wondering, there are a couple other vibrations not related to the electromagnetic spectrum that we deal with every day.

The Electromagnetic Spectrum

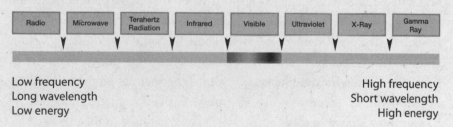

| Radio | Microwave | Terahertz Radiation | Infrared | Visible | Ultraviolet | X-Ray | Gamma Ray |

Low frequency
Long wavelength
Low energy

High frequency
Short wavelength
High energy

The physical energies we know in the electromagnetic spectrum range from radio waves to gamma rays. Only a tiny part is perceptible to us via our senses.

Sound, oddly enough, is not an audible portion of the radio wave band of frequencies. Sound is a series of compression waves that travels through matter, created by the back-and-forth vibration of an object, such as a loudspeaker. The waves are perceived when they cause a detector, like your eardrum, to vibrate. Any sound you hear as either a low or high-pitched tone is made of regular, evenly spaced waves of air or water molecules.

Temperature is also not related to the electromagnetic spectrum. The heat an object possesses is determined by how fast its molecules are moving, which in turn depends on how much energy has been put into its system. Even very cold objects have some heat energy because their atoms are still moving.

To know the mechanics of the wave is to know the entire secret of Nature.

Walter Russell

The Earth Itself Is Vibrating

Other vibrations outside you come from the earth itself. There is growing evidence that the vibrations within the earth's "body" influence your own body. It's common knowledge that the fertility of various species is timed to coordinate with the earth's seasons, tides, and cycles of light and dark, which are actually vibrations, or slow wave cycles. Dr. Joseph Kirschvink from the California Institute of Technology has found that bees, migrating birds, homing pigeons, whales, and even human beings have the capacity to synthesize crystals of magnetite, a strong magnetic mineral, in their brain tissue. These crystals, so responsive to the earth's magnetic fields, may act as internal compasses. In fact, the crystals found in human brain tissue are strikingly similar to crystals some bacteria use to distinguish up from down. Kirschvink says that whales use a magnetic sensory system, following the angles and intensity of geomagnetic fields on the ocean floor as roadmaps, and tend to beach themselves at geomagnetic anomalies. It has yet to be shown how we may be tied to the changes in the earth's magnetic fields through the magnetite in our own brain tissue.

Gregg Braden, a former computer systems designer and geologist, has done some fascinating research concerning the earth's geophysical cycles, which have repeated throughout history and been described by previous cultures. Braden studied the earth's base resonant frequency, called the Schumann Resonance, or SR. He says that for decades, the measurement was 7.8 cycles per second, which was thought to be a constant. Recent reports show the rate at 8.6 and climbing. Braden also says that while the earth's "pulse rate" is rising, her magnetic field strength is declining. According to some researchers, the field has lost up to half its intensity in the last four thousand years, and this may signify a condition leading to a magnetic pole shift. There are currently reports of magnetic anomalies, picked up by compasses, ranging up to fifteen to twenty degrees away from magnetic north. Braden, who links the earth's frequency changes to our own cellular vibration and possibly evolutionary DNA changes, says that with the weaker magnetic field and faster base frequency, old emotional and mental patterns are less locked in and we have easier access to higher states of consciousness. He says that we, like the earth, are speeding up toward a shift in energy and awareness.

We don't have to go anywhere. We are living in a global initiation chamber, with these geophysical conditions occurring on a worldwide scale. It's as if Earth herself is preparing us for the next stage of evolution.

Gregg Braden

Inside You There Is a World of Vibration, Too

So not only do you live in a world that oscillates in and out of form, where there are unlimited potential realities evolving simultaneously, where everything is vibrating and waves of energy are traveling in every direction, but inside your body, in the microcosm of your personal life, you're vibrating as well. If you place your attention inside your body, what do you notice? What's moving? The first oscillation is the wave of your breathing: draw fresh air in, transfer oxygen from lungs to blood, transfer carbon dioxide and impurities from blood back to lungs, and out with the exhale. Next, you might notice your heart pumping blood out through the arteries and back through the veins, out and back, out and back. Tune in further and the next higher vibration you find is the electrical buzz of your brain and nerves as charges are transmitted across synapses. You might sense it as tingling.

As you go deeper, you'll discover the even subtler vibration of your neurotransmitters and biochemical processes as they work with your cells. Below that, you can feel the vibration of the cells themselves. Below the cellular vibration, you might sense the vibration of molecules and atoms, and finally, the "quantum entities." By imagining this descent through the vibrations to the core place inside your body, you are moving from low frequency (breath and heart) to high frequency (molecules and atoms). As you enter any atom and are drawn into a final particle, you are likely to experience it mysteriously transforming into a wave of energy, freeing you from time and space. In that release, you'll have an experience that physics can't yet describe. It's what happens when energy becomes awareness.

Try This!

Journey Through Your Body's Vibrations

1. Sit comfortably, palms on your thighs, head level, breathing evenly. Bring your attention fully into the present moment and inside your skin. What do you notice? What's moving? Perhaps it's a sort of teeming, jiggling, wiggling, buzzing, or humming. Be with the vitality and activity of your body as it goes about its business of being alive. Feel how appreciative you are for your body and life.

2. Now notice the cycle of your breathing. Follow your breath all the way in as it brings fresh air into your lungs, then turns and flows slowly back out, taking impurities out of your body. Then let it turn at the end of the outbreath and flow back in. Stay with the movement of the wave, merging with the flow, letting it be effortless.

3. Next, pay attention to your heart beating and its pulse as it occurs in various places throughout your body. The vibration is a bit faster than that of your breath. Merge into your heartbeat, feeling the heart pump and release, pump and release.

4. Now feel the electrical buzz of your nerves. This vibration is even faster and higher in frequency than your heartbeat. Scan around your body, noticing the tingling everywhere.

5. As you go deeper, see if you can sense the vibration of the biochemical processes and neurotransmitters in your body, as important chemicals and nutrients transfer into and out of your cells.

6. Now focus on the vibration of a clump of cells anywhere in your body. Imagine viewing them under a microscope; can you feel them jiggling with their very subtle vibration? Drop into a single cell.

7. Keep going, and let yourself be drawn into a molecule that makes up the cell; here you'll enter an extremely refined vibration, that of one of the base elements, such as hydrogen or carbon, that makes up the cell.

8. Next, fall down through the molecule into an atom and feel the amazing life force contained within. You are reaching a very fast, high-frequency vibration.

9. Finally, let yourself be drawn into one of the "quantum entities," the tiny wavicles inside an atom. As you journey down into that final particle, notice when it mysteriously opens or gives way and transforms into a wave of energy and consciousness, freeing you from time and space.

10. Now you are floating in a very quiet place where there is no movement; you have spread out and are everywhere. Everything is possible. Everything is known and knowable. All you can do is *be*. As you *be*, you absorb the new guidance, instructions, and energetic blueprints that you need. You become fuller and heavier and pop back into a reality again, as a new quantum entity, a new particle in time and space.

11. Begin journeying back through the vibrations, from the fast, high frequencies to the lower, slower frequencies: atom, molecule, cell, biochemical rhythms, nerves, heartbeat, and breathing. You're back! You're rejuvenated with new life force and a new sense of purpose.

Your Brain Has Waves

Your brain is an electrochemical organ, and its electricity is measured in brain waves. There are four categories of brain waves, ranging from the fastest to the slowest. Interestingly, the fastest brain waves correspond with lower frequency awareness, while the slowest brain waves correlate with higher frequency, expanded awareness.

Beta (13 – 40 Hz, or cycles per second) Beta, the fastest of the brain waves, is associated with an alert waking state, when your brain is aroused and engaged in mental activity. When you read in bed before going to sleep, you're probably in low beta. The more intense the activity and arousal, as with fear, anger, hunger, or surprise, the faster the frequency.

Alpha (8 – 13 Hz) Alpha brain waves are slower and appear when you're relaxed but not drowsy, with an effortless alertness. They are found during tranquil states, such as light meditation, reflection, daydreaming, biofeedback and body-mind integration, light hypnosis, creative visualization, artistic and intuitive processes, time in nature, rest, and exercise.

Theta (4 – 8 Hz) Theta brain waves are much slower and are associated with drowsiness, the first stage of sleep, dreaming, deeper levels of meditation, inspired creativity and imagination, increased recall, and mystical states of intuitive perception. Theta can feel trancelike, as when you're driving on a freeway or taking a long shower, lose track of time, and perhaps find a free flow of ideas or visions surfacing in your mind.

Delta (½ – 4 Hz) Delta brain waves are very slow and are found during deep sleep. They are connected with sleepwalking and sleeptalking, as well as deep trance and processes of self-healing.

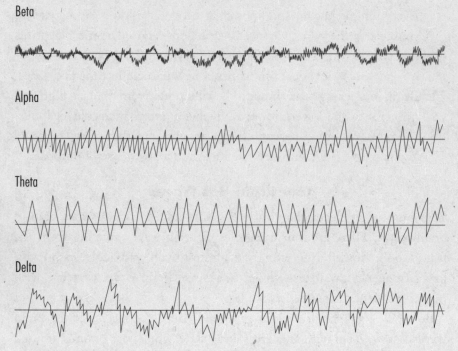

Beta

Alpha

Theta

Delta

Your brain shifts between four basic frequencies of consciousness. At each vibratory level, you function differently; these phases are evident during your sleep cycle.

At night, you rotate through several phases of sleep. In the early stage, your brain waves slow from beta to the more relaxed alpha state, and imaginative pictures might drift through your mind. Your muscles relax, and your pulse, blood pressure, and temperature drop. Next, your brain waves slow to theta level. You're now in a light sleep state characterized by many bursts of brain activity. Most dreams occur during a state called *REM* (rapid eye movement) *sleep*. During REM sleep, brain waves increase from theta and temporarily include a mix of faster frequencies closer to your waking state. If you're awakened during this period, you'll easily remember your dreams. In the last phase, your brain waves slow even more, reaching delta, where you experience deep, dreamless sleep. If awakened, you'll feel fuzzy and lost, resist waking fully, and drop back to sleep almost immediately. Interestingly, our hearts show electrical vibration patterns in almost the same range as the brain's delta waves.

Your Brain Waves Correlate with Your Awareness Levels

Next, I want to make a leap, one that science has not made yet but which is quite evident to practitioners of body-mind-spirit disciplines: namely, that the energy frequencies of matter have matching consciousness frequencies. We see this showing up in biofeedback training, when different brain waves produce unique experiences and ways of knowing. When you become familiar with the states of consciousness that result from various brain wave activities, you'll notice that the fast *beta* waves correspond to the shallow consciousness of daily reality and the busy "linear" mind. The more hyper and contracted your mind is, the less far-reaching your awareness is. Some say the ego is a function of this consciousness level. Biofeedback research shows that as brain waves slow to *alpha*, you become less worried, more open, and aware of more subtle kinds of information. You have access to deeper regions of memory, symbolism, and insight. In fact, you're now able to focus on what's been suppressed and stored in your subconscious mind without the dread common in the normal waking state.

As your brain waves slow further to *theta*, you find understanding about the nature of your true self. The ego begins to "die" and be replaced by soul awareness. It's been shown that when the mind turns inward and focuses on self-reflection and inner processes, your brain waves shift from beta to alpha to theta, especially as outside stimuli are shut out. Meditators often drop into deep theta states and report that they feel unity within themselves and with all other beings. It's ironic that when viewed from our normal waking consciousness point of view, the deeper states appear sleeplike and trancelike, yet when we become conscious within them, a much more expansive awareness awakens.

Entering the *delta* state brings experiences of being out of the body; your sense of self expands to become collective and universal. There is no time or space, and you can easily shift into other dimensions of awareness. It's not easy to achieve these deep states while alert and conscious, as delta takes you into a unified, nonindividualized awareness that is overwhelming to the ego—and falling asleep instead of exploring the experience is often a comfortable "escape." Delta might be thought of as physics' many-worlds quantum reality in which all parallel worlds coexist and evolve simultaneously.

Our normal waking consciousness, rational consciousness as we call it, is but one special type of consciousness, whilst all about it, parted from it by the filmiest of screens, there lie potential forms of consciousness entirely different.

William James

A few interesting facts come to mind here. Researchers have found that the brains of people who suffer from ADHD are primarily functioning at theta level. They make progress when trained through biofeedback to develop the beta state. Most of us are overstimulated with beta (especially with the escalation of the Information Age), trying to slow down and function more from an alpha state, while hyperactive kids, for example, are actually trying to speed up, live inside their bodies, deal with ordinary reality, and think straight and logically. It may be that they're naturally tuned to higher states of consciousness, and like ET (the extraterrestrial), need to learn how to function in the mental and emotional atmosphere on earth.

Second, it's been found that when your brain functions at the slower frequencies of alpha, theta, and delta, it produces more beneficial neuropeptides and hormones, such as endorphins, serotonin, acetylcholine, and vasopressin, which help relieve stress and pain and increase learning and memory. Might memory loss and Alzheimer's disease, which are linked to low acetylcholine levels, be in any remote way related to too much busy beta consciousness and not enough time spent developing the more expanded awareness that corresponds with alpha, theta, and delta brain waves?

Consciousness, Like Energy, Has Its Own Spectrum

We've seen how brain waves tie to different levels of awareness, but might there be an even more detailed breakdown of specific kinds of perception in each brain wave frequency? Christopher Lenz is a longtime facilitator of the Virginia-based Monroe Institute's Hemi-Sync consciousness expansion training, which uses sound patterns that can have dramatic effects on brain waves and states of consciousness. He described to me the levels of awareness, particularly those that occur in theta and delta states, that have been visited and defined by participants in the training over the years.

In the training, specific frequencies of continually changing chords and beats, sent binaurally through stereo headphones, transport participants from ordinary to nonordinary states of consciousness. They travel from profound relaxation, then expand beyond sense perception, then move beyond time and space and enter a state that bridges them into nonphysical realities. Lenz reports that to help make sense of it all, an overall spectrum of awareness that divides consciousness into forty-nine levels and seven octaves was delineated for the institute by a nonphysical, channeled entity called Miranon. Training participants then visited most of these levels and reported back, building a consensus based on the commonalities in what they experienced.

In Miranon's system, **Levels 1 through 7** pertain to the consciousness of the plant world. **Levels 8 through 14** correspond to the animal kingdom. **Levels 15 through 21** relate to the kinds of awareness in the human realm. Lenz said a number of people have found that the human frequencies loosely parallel the functions of the seven *chakras*—the energy vortices or centers in our body. Chakras might be thought of as points where higher energies are transferred into the physical world via an intermediary energy or light body, composed of subtle energy often called *chi*, or *etheric energy*. In general:

Level 15 correlates with the first chakra at the base of the spine, which focuses on survival and endurance, the will to live, maintaining and circulating vital energy, keeping us interconnected with the earth, and supplying us with the earth's core energy.

Level 16 corresponds to the second chakra in the lower abdomen, which focuses on sympathy and the ability to feel emotions and energy fields. It helps us connect and relate with others and the world, influences sensuality, desire, hunger, and sexuality, and promotes creative and reproductive urges.

Level 17 connects with the third chakra at the solar plexus, which focuses on personal will, personal power, self-control, control over the outside world, and attraction-repulsion and fight-or-flight responses to incoming stimuli.

Level 18 relates to the fourth chakra, the heart center, which focuses on the interconnection of our physical and nonphysical selves, the balance of living in alignment with universal principles, and the perfect understanding of our interrelatedness with others and nature. It promotes compassion, refined empathy, and trust.

Level 19 corresponds to the fifth chakra at the throat, which focuses on the transition from personal will to higher will, faith, connection to higher sources of guidance and information, inspired self-expression, and authentic communication.

Level 20 correlates with the sixth chakra, or third eye, in the central brow area, which focuses on the ability to see through form into underlying energy patterns and dynamics that connect us to nonphysical experiences of self. It opens our intuition, imagination, inspired creativity, and finely tuned awareness.

Level 21 corresponds to the seventh chakra, or crown, which focuses on the transition of our identities from individual ego to soul, linking us to the highest sources of guidance, cosmic understanding, and interconnection with all other souls and the universe. This is often experienced as a meeting place with nonphysical beings.

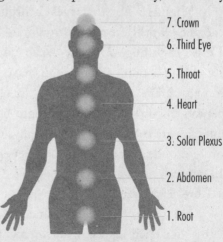

7. Crown

6. Third Eye

5. Throat

4. Heart

3. Solar Plexus

2. Abdomen

1. Root

The Chakra System

There Are Awarenesses Beyond the Human Realm

Lenz continued, describing levels 22 through 28, the realm beyond normal human awareness, which focuses largely on after-death experiences. I find these levels particularly interesting, since they are a mapping of realms that I, and many other clairvoyants and mystics, have visited in altered states of awareness.

Level 22 correlates with higher-level dream realities, dementia, delirium, and being anesthetized. It is found with coma patients and can be used to make contact with them. It's also where people who've died and don't believe in an afterlife may "sleep" for an indefinite time.

Level 23 is associated with people who've had a fear of death and don't know they've died, or who can't get past a limiting idea or emotion after death. This can be typical of suicides, addicts, or people who die with high levels of resistance, bitterness, or grief.

Levels 24, 25, and 26 pertain to levels of fixed beliefs and expectations concerning people's concepts of what the after-life will be like. For example, if someone believes they'll hear trumpets blow when they reach heaven, they do. Or if they believe in a religious figure, they'll find him or her. The levels also pertain to ironclad belief systems and worldviews people hold about how life on earth works. If the family is of prime importance, they will remain in their familial patterns after death. If sexual freedom or power over others is important, they'll continue to act that way. Lenz says people can stay at these levels for long periods—centuries, in fact—until they discover other possibilities with the help of guides.

Level 27 is a realm known as "The Park." It's a kind of reception center where most people go after death. To many, it appears as a huge, green, tranquil park with gorgeous trees and lawns. Here, people reconnect with loved ones who've died, work with guides to understand what they were learning in their lives, go to a rejuvenation center to recover their energy, look at their possible future lives, learn in the vast Akashic Records library where planetary memory is stored, or just relax and play until they're ready to move on. Some people are trained here as rescue workers and guides who learn to drop back into lower levels to bring "lost" souls up to the state where they can be free to move forward.

Level 28 is a more abstract state that is the last association with being human, sometimes referred to as "seventh heaven." Here, people connect with high teachers, saviors, and prophets, and it is where nonreincarnating souls who are integrating all their lives abide.

Moving beyond into the next three octaves, Lenz said that most are not well documented. Levels 34 and 35, however, are called "the area of the Gathering," where we meet extremely high impersonal intelligences who have gathered around the earth to witness two important events that will occur on earth simultaneously, which they deem to be a rare occurrence. They want to help us. The events have not been clearly delineated, however, by the people who reached these levels of awareness. Level 49 pertains to the underlying workings of astronomy and understanding the great mysteries of the cosmos and galaxies.

Hopefully, you're beginning to get a glimpse of the many interpenetrating spectrums of energy and awareness in both your external and internal realities. The Monroe Institute's description of awareness levels is just one of many hierarchies of consciousness. Almost every religion and metaphysical system has its own version of what's there to be discovered in our slower, deeper brain wave states. Pioneers throughout history have explored higher states of awareness and energy, mapping the territory using the filters of their own cultures, yet in spite of small differences, there is great similarity among the various models. You, too, will eventually be able to intentionally raise your own frequency to levels where higher-dimensional knowing will seem normal.

> The cosmos is the ordering of number. Perception is the imaging of form contained in the potential of number.
>
> Robert Lawlor

What State of Awareness Corresponds with 214?

I am reminded of a dream I had many years ago where I was being quizzed for what seemed like hours by a panel of spiritual examiners, as though I were taking orals for an advanced degree. They would call out to me: "What physical sensation matches one hundred twenty-nine?" "Itching!" I'd call back. "What emotion corresponds with two hundred fourteen?" "Grief!" I'd say. "What physical state correlates with thirteen?" "Paralysis!" I'd answer. "What state of mind equates with five hundred twenty-five?" "Enthusiasm!" I'd reply. I woke up with the sensation of a scale with finely differentiated hairline markings inside me, as though I were made of layers of thin tissue paper stacked up in every direction.

Later, I began to study numerology, and after many subsequent years of working with the number combinations that result from a client's birth date, I learned how numbers indeed correspond to states of awareness, life lessons, personality traits, and particular kinds of events. Numbers are frequencies. And every frequency reveals a world or reality with a unique kind of knowledge and particular rules by which it functions. By tuning your sensitivity to different numbers, you can discover new worlds of information and train yourself in a basic energy skill: *traversing the scales*.

Try This!
Realms of Number Vibrations

1. Sit comfortably, palms on your thighs, head level, breathing evenly. Bring your attention fully into the present moment and inside your skin.
2. Imagine you are in a glass elevator resting at ground level, and through the door you can see a large number 1 on the wall. This is your normal, everyday reality.
3. Notice that this elevator has the ability to go to floors 1 through 9. You are going to move through the various levels and come to one you'd like to experience.
4. Close your eyes and count very slowly in your mind, noticing any attraction, no matter how mild, to a particular number: 1 – 2 – 3 – 4 – 5 – 6 – 7 – 8 – 9, and back, 9 – 8 – 7– 6 – 5 – 4 – 3 – 2 – 1, then back up again. See if you can sense the different vibrations as you move through the numbers. Finally, you will be drawn intuitively to a particular number. The elevator will come to rest at that level. As the elevator door opens, step out.

5. There is a new world here, resonating to the number you've chosen. Feel the vibration and let yourself adjust to it. This is a frequency you need for your body, emotions, or mind, and it contains a particular kind of energy and wisdom, and can facilitate abilities and skills you may need. Wander around; what do you see, hear, smell, or feel? What emotions and feelings are triggered? Absorb deeply. What thoughts and topics do you start thinking about? Do insights come to mind about something you need to do?

6. A guide now comes to greet you. The guide is a spokesperson for this level, and he or she will begin a dialogue with you, giving you information and advice you need. In your imagination, let the information flow. If you want to write using the direct writing technique described in *To the Reader*, continue to stay focused and gently begin writing whatever guidance comes to mind.

7. When you have received enough, thank the guide and step back on the elevator. You may either go to another level and repeat the process there or return to your normal waking consciousness at level 1, feeling the steps in vibration as you go.

Your Body, Emotions, and Thoughts Influence Your Frequency

Everyday reality presents you with yet another array of vibrations. These are the ones that pertain to the state of your physical body, emotions, and thoughts. In an ordinary day, you move freely up and down the scales of comfort-discomfort, action-inaction, happiness-displeasure, and logic-inspiration, to name a few. The following chart gives you a general feel for the wide range of vibrations you might experience. The columns do not correspond to each other horizontally.

Here's an example of how personal vibrations might change over the course of a day: perhaps you wake up with great enthusiasm to get started on a new sculpture you've envisioned, and you're hungry for the tactile experience. As you answer your email first thing, a snafu with your computer frustrates you, and you dip toward anger. Then a friend calls and invites you for dinner on the weekend, and you bounce back toward gratitude. You begin work on your sculpture and inadvertently injure yourself

The Scales of Everyday Vibrations

HIGH FREQUENCY fast/ expansive/ soul/love	BODY	SENSES	EMOTIONS	THOUGHTS
	Full Presence	Communion	Love/Empathy	Wisdom/Oneness
	Perfect Health	Direct Experience	Generosity	Direct Knowing
	Joyful Movement	Ultrasensivity	Joy/Gratitude	Inspiration/Insight
	Flexibility	Intuition	Enthusiasm	Fluid Creativity
	Responsiveness	Clairvoyance	Desire/Motivation	Discovery/ Exploration
	Comfort/Rest	Clairaudience	Pleasure	Receptivity/ Openness
	Exhaustion	Clairsentience	Sincerity	Boredom/ Impatience
	Tension/Stress	Vision	Contentment/ Trust	Distraction/ Absence
	Periodic Pain	Hearing	Disappointment	Projection/Blame
	Chronic Pain	Touch	Frustration	Logic/Proof
	Addiction	Taste	Doubt/Insecurity	Beliefs/ Control Games
	Disease/Illness	Smell	Fear/Panic	Obsession
	Trauma/Injury	Attraction/ Repulsion	Hate/Rage/ Refusal	Overwhelm
LOW FREQUENCY slow/dense ego/fear	Loss of Function	Gut Instinct	Guilt/Shame	Psychosis/ Neurosis
	Paralysis/Coma	Subconscious Reaction	Depression/ Apathy	Suicidal

with a chisel. Your body experiences pain and your hand throbs for the rest of the day. That distracts you, and you feel drained.

You take a break, do some chores, and get bored. You decide you can't work and turn on the television. You enjoy the visual stimulation of a beautiful movie but are agitated by the loud, rapid-fire commercials. It's time for bed, and as you try to fall asleep, you're plagued with worries about money and being good enough at what you do. You realize you're too overstimulated to sleep, so you read a book that transports your thoughts to more spiritual matters. This calms you, and you decide to meditate, which reassures you. Now you feel peaceful and reconnected with your wonderful, talented, artist self, and it's easy to fall asleep. Your dreams are creative and dynamic.

Try This!
Track Your Daily Vibrations

- Think back over your day so far. What physical, emotional, and mental state did you greet the day with?
- How did you move through the frequencies as the day progressed?
- What states are you living in right now?
- When you recall a typical day, is there a pattern of energy states you tend to repeat without realizing it? What is your typical early-morning vibration? Mid-morning? Early afternoon? Late afternoon? Early evening? Nighttime?
- Think back over your week. What were the predominant vibrations you lived in?
- What frequency levels have you been dreaming in lately?

You Have Your Very Own Personal Vibration

Since life itself oscillates in and out of form, and your mind alternates between being conscious and unconscious, when it comes to daily experience, it's normal to shift from the low- to the high-frequency end of the body-emotion-thought scale and back down again. Many people I talk to are trying to find a permanent way to remain in the "good stuff," be happy all the time, and never drift off purpose. But we are vibrating beings and we move through our experiences by virtue of life's wavelike nature. You, as well as all other people, have your own *personal vibration*—and it changes depending on what you're thinking, feeling, and doing. The most important thing to know is that you can influence it. You can let your personal vibration match the world's chaotic soup of frequencies and feel buffeted about helplessly, or you can determine how you want to feel. When you choose to attune to the frequency of your soul, your personal vibration stabilizes at your "home frequency," which we'll discuss in detail in chapter 5. You may still bob up and down a bit, but it is much easier to maintain your clarity when you know how to intentionally focus your personal vibration. But what exactly is your personal vibration and how does it function?

1. Your personal vibration is the overall vibration that radiates from you in any given moment. That frequency is a combination of any of the various contracted or expanded states of your body, emotions, and thoughts outlined in the previous chart. It naturally fluctuates. One minute you may experience a combination of physical pain, emotional victimhood, and mental overwhelm. The next, you may be engaged in fluid physical movement that brings emotional pleasure and an interest in exploration. If you're upset, you might use gut instinct and react, and if you're calm and happy, you may have highly intuitive insights.

2. The vibration of one aspect of your makeup affects the vibrations of the other aspects. If you've been depressed for a while, the low emotional state may cause your imagination to dry up and your thoughts to be more hopeless and negative. The lack of warmth and enthusiasm might also cause your body to be more stagnant and sluggish. On the other hand, if you move your body regularly, say, by walking or dancing, your emotions will be spontaneous, and you'll experience greater fluidity of thought as well. What's great about this is that you can improve your overall personal vibration by improving one part of your makeup—just move up the scale to a quality that feels better. For example, if you're exhausted, move into comfort and rest. If you're frustrated, try shifting to contentment. If you're obsessed, try looking at your underlying beliefs. Start anywhere and the rest of yourself will follow.

> Every change in the physiological state is accompanied by an appropriate change in the mental-emotional state, conscious or unconscious, and conversely, every change in the mental-emotional state, conscious or unconscious, is accompanied by an appropriate change in the physiological state.
>
> Elmer Green

3. Your personal vibration is affected by vibrations in the world and in other people. Because bodies resonate like tuning forks, you can come in contact with another person who is negative and nervous, or earthy and calm, and your body will copy theirs. Maybe you're flowing along cheerfully and a friend calls who says she has a problem that's your fault. It's easy to pick up her vibration and start blaming her in return for something *she's* done. Before you know it, you feel upset and drained. Or perhaps you wake up tired and become more lackluster during your boring commute in heavy traffic. Your brain doesn't want to work until an excited coworker bursts into your office with great news. His energy is upbeat and motivated. Soon you're in a creative frame of mind and highly productive.

Some people are "uppers" to be around, and some are "downers." Similarly, locations carry vibrations—one restaurant may feel "creepy" while another makes you feel safe and cozy. Some cities stimulate your creativity and social nature while others isolate you and make you feel hopeless. You might try being the one who "sets the tone" for others in any given situation, because your personal vibration affects others as readily as theirs does you.

4. Your personal vibration is generated from inside you by your own choices. It's really up to you how you want to feel. You have a natural vibration—your *home frequency*—that's the way your soul feels, the kind of gorgeous, bright frequency that babies radiate. This is covered over with emotional and mental clutter as you go through life. You can uncover it and shine out any time you want. One of your choices, usually quite unconscious, may be to get attention and feel good by identifying yourself as a victim and acting as though you've been wronged or that you *are* wrong. This unconscious choice keeps you operating at a low personal vibration. To the extent that you hold these ideas, you'll allow other people, places, and events with lower vibrations than yours to bring you down. Or you'll assign people with naturally high vibrations the responsibility for making you feel better.

5. Your personal vibration improves dramatically the more you allow your soul to take charge of your life. When your personal vibration is the result of limited thinking, negative emotions, or the need to control yourself and the world, you're not going to have a very good time here on earth! As you move through the transformation process, those contracted feelings and beliefs rise from your subconscious mind to be cleared, and a new perception based on your soul's expansive, loving wisdom opens to you.

In the next chapters, we'll delve more deeply into how you can develop healthy sensitivity and keep your personal vibration at a level that optimizes the way you function in life, brings you success, and allows you to be naturally happy and entertained.

> The greatest revolution in our generation is that of human beings,
> who by changing the inner attitudes of their minds,
> can change the outer aspects of their lives.
>
> Marilyn Ferguson

Just to Recap . . .

You are surrounded by many energy vibrations in the outside world, from the eighty-one octaves of electromagnetic frequencies, to the vibrations of sound and heat, to the earth's Schumann Resonance, to the basic wavicles of matter. Some are perceptible through your senses, but most are not. In your inner world, you're alive with vibrating waves of energy and consciousness, from the cycles of your breath and heartbeat to your electrical brain waves. Within your brain wave levels are a variety of awarenesses, from the seven kinds of chakric consciousness to awareness of after-death experiences and other dimensions. You can traverse the frequency scales with your sensitivity and discover new worlds and knowledge in each frequency.

Your personal vibration or energy state is a blend of the contracted or expanded frequencies of your body, emotions, and thoughts at any given moment. The more you allow your soul to shine through you, the higher your personal vibration will be. Your personal vibration is affected by other people's vibrations and the vibrations of the world, yet ultimately, how you want to feel is your choice.

Home Frequency Message

As I explain on page xxi in *To the Reader*, I've included these pieces of inspired writing at the end of each chapter as a way for you to shift from your normal, speedier reading mind to a deeper kind of direct experience. Through these messages, you can intentionally change your personal vibration.

The following message is meant to transport you into a way of knowing the world that's close to the way you'll experience life in the Intuition Age. To move into the *home frequency message*, just downshift to a slower, less hurried pace. Take a slow breath in, then out, and be as calm and still as possible. Let your mind be soft and receptive. Open your intuition and prepare to *feel into* the language. See if you can experience the deeper realities and feeling states that come alive *as you read*.

Your experience may take on greater dimension in direct proportion to the amount of attention you invest in the phrases. Focus on the words a few at a time, pause at the punctuation marks, and "be with" the intelligence delivering the message—live and right now—to you. You might speak the words aloud, or close your eyes and have someone else read them to you and see what effect they have on you.

MOVING SMOOTHLY THROUGH THE VIBRATIONS

Listen to the world: sounds move from sources to ears. Dogs turn heads toward intruders, dolphins move to schools of fish, music pours from instruments, stories flow from the mouths of teachers. There are: insect-clicking, thunder-booming, mother-child whispering, groaning from effort, screaming from loss, shouting for victory. Listen to the layers of vibration, feel them as tangible, impressing you, making you resonate.

You are the wind: now breeze, now gust, now storm. You are the light: now white, now rainbow, now black, now diamond and transparent. You are the wave of life: the going out, the coming back, the ballooning to unknown full-nesses, the disappearing into unknown pinholes. You don't want to stop, and you do want to stop. You think the rocking is you. You know the rocking is not all of you.

Under surface vibrations, there is something inviting, something renewing, something supremely effective, something frightening at first. Below the waves is the quiet-where-movement-slows. No need to talk: think a thought to someone and they think it, too. Imagine a gift appearing in a friend's hands and it is there, no commotion of creation. Knowing and doing without so many waves. Can you feel how calm, how easy? The calmer you are, the faster and more accurately your intention translates.

Below this place of calm translation, another place. No movement at all. Here is something pure: awareness, silence, profound peace, permeating, freeing love, total understanding that surpasses and relieves the mind. No directionality, no force, no need. All acts of creation begin here and end here. Here you learn by being the universe. In any moment, you can descend through the frequencies into stillness. Your parts unify. Time stops and you think, Nothing is there. Then suddenly, a "laugh" explodes you to vibrating bits again: everything is here! You rise from the purity to feel a new motive and choose a new vibration and flow somewhere new and know the joy of that until the stillness calls you back once more.

The deepest state is the highest; the calmest is the most accelerated. The most intimate, loving, connected feelings are the most effective. Love is the most motivating and creative frequency of all. Truth is love moving through your mind. Harmony is love's resonance coordinating and attuning all life's vibrations. You-the-mind love vibrations because Mind is made from vibrations, yet you-the-soul love stillness because Soul is made of indivisible presence and unmovable love. Have a feast of frequencies, digest quietly, use your presence to create with frequencies, and appreciate quietly.

3

Becoming Aware of Your Feeling Habits

Emotions are not "bad." At the roots of our emotions are primal energies
which can be put to fruitful use. Indeed . . . the *energy* of enlightenment
arises from the very same natural origins as those which
give rise to our everyday passions and emotions.

James H. Austin, MD

Irecently counseled a teenage girl who was interested in intuition. In
spite of the piercings in her lips and her black eye makeup, Megan's eyes
glowed with beauty and she was incredibly open-minded. She shrugged
when I asked her what kinds of questions she had, but about halfway into
the session, she thought of something. "Sometimes I get this sudden feeling that everything is going to go wrong, like the good stuff is going to get
wrecked. What causes that? Because I'm always afraid my boyfriend is
messing around with other girls—but he's not—and I think I'm hurting
our relationship."

I felt into her body and could sense the place in her solar plexus where
the sudden feeling of dread would typically hit her, seemingly out of
nowhere. Attached to the sensation I had a vague image of her father.
I asked if her father had left her and her mother. "Yes," she said, "when I was
little." I could feel how her tiny body had reacted with that original contraction of fear, that confusing sense of abandonment and consternation

about whether she had somehow caused her father to leave. It was as though some vigilant part of her mind was awakened then that hadn't needed to be there before, a part that was going to be on the lookout from now on for situations where a similar sort of devastating surprise might happen. "I'm never going to be caught off guard again!" it was saying.

I could feel how Megan's natural tendency was to trust and want to love, and when she did, her subconscious mind relaxed. Then suddenly, it would remember: "There's something familiar about this situation—you could be abandoned: *Watch out!*" And it would contract, out of the blue. Instead of realizing she was reliving a past experience, Megan was projecting the anxiety onto whomever was nearby, and she had already lost several friends and a previous boyfriend because of it. I worked with her to help her understand that her body was ultrasensitive and aware, that it was simply fixated on that first shock, and that she could easily reprogram the response, especially because she was still young and so close to the original experience— she hadn't lived through twenty more similar losses caused by her vigilante mind revalidating the original contraction, as so many adults have.

You Were Born, and Still Are, Vibrationally Sensitive

Like Megan, you came into this life wide open and trusting, expecting joy and a free exchange of nurturing love. Certainly, we are all born empathic and sensitive to vibrations. In *Evolution's End*, Joseph Chilton Pearce describes how, a few days after conception, the embryo forms a clump of vibrating cells that becomes the new heart. These cells seem to be sound sensitive and tune to the mother's heartbeat and breath, which appear to be necessary for further forming the infant's heart. The mother's emotional state, and any repetitive patterns of behavior, are imprinted on the fetus hormonally and through the tone of her voice. After the infant is born and its body is still loaded with adrenal hormones, the mother's instinct—and the father's as well—is to hold the newborn to the left, close to the heart. The parents' hearts stimulate the baby's heart, which activates the brain-mind and reassures the child that it is safe. It is common knowledge that playing a recorded heartbeat can reduce babies' crying by 40 to 50 percent.

Pearce goes on to describe how newborns still live in a subtle world, almost completely connected to the mother's body. He says, "When the mother's subtle sphere overlaps the infant's, a major communication takes place. This subtle communication might be below the level of awareness for the mother, but it is the only level of awareness fully active in the infant." So, the heart-to-heart skill of *feeling into*—child to mother and mother to child—is our original method of growth and survival.

In his book *Emotional Intelligence*, Daniel Goleman describes how infants and young children commonly display empathic connections to each other. "Virtually from the day they are born, infants are upset when they hear another infant crying," Goleman reports, and they "react to a disturbance in those around them as though it were their own, crying when they see another child's tears." Children often imitate the distress they see in another child. For example, when another child hurts her fingers, a one-year-old might put her own fingers in her mouth to see if she hurts, too. A toddler might try to give a baby a toy when it cries or stroke its head. In later childhood, Goleman says that children begin to understand that someone's distress can be related to their situation in life, and they can feel sympathetic toward an entire group, like the poor or the outcast. Vibrational sensitivity and empathy, it seems, are fundamental human tendencies, with roots going back to the womb.

> We must all learn to listen to one another with understanding and compassion, to hear what is being felt by the other.
>
> Thich Nhat Hanh

Psychologist Elaine Aron, in her book *The Highly Sensitive Person*, says that "highly sensitive" adults make up approximately 15 to 20 percent of the population. These are people who sense things that are unobserved by others, have a deep understanding of any environment and its problems and potentials, and often need more space and time to process the amount of input they absorb. HSPs are easily overstimulated to painful and paralyzing levels, and are often naturally introverted, intuitive, visionary, and involved with their soul and spiritual life. Carl Jung wrote that highly sensitive people are more influenced by the unconscious mind, which gives them access to important information that can contain prophetic insight. According to Aron's research, the "moderately "sensitive" type accounts for another 30 percent of the people she queried. She says the remaining 50 percent think of themselves as being not sensitive or "not at all sensitive."

Perhaps we might all be more consciously sensitive and empathic if our culture weren't so analytical, materialistic, and competitive. Perhaps, in part, we're trained out of it; our schools emphasize math and computer skills over art and literature, sports over dance and music appreciation. Then again, early wounding from dysfunctional families, as in Megan's case, can shut us down and force us into our heads, into compensating behaviors, or even out of our bodies into dissociative states. Yet given our purely vibrational roots, it seems unbelievable that we can become insensitive to knowing others via vibration and resonance, especially now when the frequencies inside and outside us are accelerating. I suspect that a much greater percentage of people are making their way into the Highly Sensitive Person category these days, as stoicism breaks down and the unhealthy feeling habits we've developed to protect ourselves—like addictions, cocooning, or aggression, to name a few—no longer work.

Are Women More Sensitive Than Men?

Elaine Aron found that by school age, most highly sensitive people had become introverts, but girls were allowed more freedom to express emotion and could even be out of control or helpless, while boys had to be more stoic to survive. Daniel Goleman summarizes the role of gender in emotional and empathic ability when he says, ". . . there are far more similarities than differences. Some men are as empathic as the most interpersonally sensitive women, while some women are every bit as able to withstand stress as the most emotionally resilient men. Indeed, on average, looking at the overall ratings for men and women, the strengths and weaknesses average out, so that in terms of total emotional intelligence, there are no sex differences."

It is commonly assumed that women are more intuitive than men. It is true that because women have more fibers connecting the two halves of the brain, they have a more bilateral, or both-sides-at-once, way of perceiving. Men tend to perceive unilaterally, one side at a time. For this reason women find it difficult to feel separate from the environment and other people, and they receive much subtle information through that relatedness. Men can be just as intuitive, but they need to make a conscious choice to shift into an intuitive mode, since compartmentalized perception is more natural to them. This difference may pertain to our natural tendencies toward sensitivity as well.

How Sensitive Are You?

It is my observation that we are all equally sensitive—or we wouldn't be alive. Do you realize how many times you've instinctively steered clear of danger or experiences that weren't right for you, and how many times you've accurately chosen what furthers you? Sure, you've made mistakes, but because you're sensitive, you've learned from them. I prefer to reframe sensitivity so that instead of being highly sensitive, moderately sensitive, or insensitive, we think of ourselves as having degrees of consciousness about our sensitivity. Our sensitivity is always working, but we are relatively aware or unaware of it.

> All great discoveries are made by men
> whose feelings run ahead of their thinking.
>
> Rev. Charles H. Parkhurst

Many of us are unconscious or unaware sensitives, picking up quantities of subliminal information and ideas that remain just below the surface of our conscious minds without knowing what we're doing. When this happens we experience pressure from what is seeking to be known, and this can result in feeling overwhelmed or paralyzed, or upset and irritable, seemingly for no reason.

Some of us are conscious sensitives, able to identify and use the subtle data we receive via feeling from the world around us, without negative repercussions. We sense right information, timing, action, and speech and are successful in daily life.

Still others are becoming highly conscious sensitives, where you work successfully in the world with high levels of sensitivity and also feel into spiritual dimensions to gain wisdom, interact with nonphysical beings, and intentionally materialize soul-based creations.

You are probably a mix of all three categories, unconscious about certain sensitivities, alert and intentional about others, and stretching into new higher territory at times. If you put your attention on ways you've gained insight and taken action based on your sensitivity to vibrations, you'll probably find you use the ability quite often.

Try This!
How Have You Experienced Sensitivity?

1. In your journal, describe the components of three positive or negative experiences you've had where sensitivity was the predominant mode of knowing, acting, or communicating. For example: "I apologized to Sue because I could feel how hurt she was by my previous comment," or "I watched a documentary about polar bears dying and couldn't function for the rest of the day," or "I could tell the meeting was going to degenerate into a gossip session." What did it feel like before and after, emotionally and physically? How did you realize what you were feeling? How did other people respond?

2. Now describe the components of three experiences during which other people were sensitive or insensitive toward you or someone else. How exactly did it affect you?

3. This week, pay close attention to situations in which you could be more sensitive, and open your emotional and body awareness to see how you might know more, contribute more, or help others feel more validated and understood. Notice how you feel after you've applied heightened sensitivity in a practical situation.

You Have Both Healthy and Unhealthy Feeling Habits

From the time you were born, you began developing a subliminal system for feeling your way through life. Depending on the reception you received, your highly sensitive little body learned to either stay open and radiant or contract in self-protection. In time, these responses crystallized into unspoken rules for survival. They became *feeling habits* that kept you alive while allowing as much of your soul to shine through as your baby mind felt was safe.

Some of the feeling habits are healthy ones, like withdrawing inside yourself to see what feels right for you in each situation, or learning to copy whole patterns embodied by people you want to learn from. But many of the feeling habits you developed were unhealthy, designed to protect you or reinforce misperceptions your baby mind made about the inhospitable nature of the world. Perhaps you learned to "leave your body" and become insensitive to the terrible pain of your parents fighting or hitting you. Maybe you became ultrasensitive to your sad mother's bouts of depression so you could merge with her to make her happy, so she'd keep taking care of you. Today these feeling habits are second nature, and the unhealthy ones keep your personal vibration low and act as obstacles to your transformation.

To shed some light on how, what, and how much you let yourself feel, let's explore some of your feeling habits. Your healthy feeling habits help you access information, make good decisions, and improve your relationships. Your unhealthy feeling habits point to underlying emotional wounds and blocked energy flows and show where you have potential to develop new, healthier sensitivity.

Vibrational Sensitivity Survey

This survey is meant to help you understand your feeling habits. There is rating, but no final scoring, involved. Rate the following statements from 1 to 10, with 10 indicating a high incidence or the most truth. Review your results, looking for your particular pattern of sensitivity and your ideas about becoming more sensitive.

High Scores = Healthy Feeling Habits

1. I trust my gut instinct about new people and ideas. _____
2. I easily sense truth and lies. _____
3. I immediately discern other people's moods. _____
4. I trust my own good ideas and guidance. _____
5. I trust that I can materialize and have what I want. _____
6. I trust that my emotions bring me useful information. _____
7. I am immediately aware of positive or negative vibrations in spaces that I enter. _____

8. I can feel nonphysical beings, energy fields, and people's souls. _____
9. I often know what other people are thinking. _____
10. I know the right moment to take action. _____
11. I notice vibrational information when I'm in a neutral or helpful mode. _____
12. When I am overstimulated, I create more space and recenter in myself. _____

High Scores = Unhealthy Feeling Habits

1. I am upset by my own and others' pain. _____
2. I have a low tolerance for reading about or watching violence. _____
3. I am bothered by intense stimuli and chaos. _____
4. I get frazzled when I have to do too many things at once. _____
5. Showing my emotions makes others distrust me. _____
6. Making decisions based on feelings leads to failure. _____
7. Too much news exposure or the general public paralyzes me. _____
8. When I am overstimulated, I escape into "mindless activities" or addictions. _____
9. I notice vibrational information when I'm in a defensive or critical mode. _____
10. If I were more sensitive, I wouldn't be able to properly function in the world. _____
11. If I were more sensitive, I might go crazy or develop problematic mental issues. _____
12. If I were more sensitive, I might get seriously sick. _____

After examining your answers, you may notice that there are specific areas where you are largely oblivious or very much in touch with how you feel. You may find you have a high or low tolerance for vibrational stimulation, that you shift into negative perception easily or not so easily, or that your beliefs make it easy or difficult to expand your sensitivity. This is good information—not right or wrong, but a place to start being more intentional with your perception. You might want to write about your current pattern. Where does fear influence you the most? What might you do to shift your unhealthy habits toward the healthy side? Where are you already skilled in your ability to trust and use vibrational information? Where would you like to develop healthier feeling habits?

How Your Feeling Habits Develop

Let's take a little journey back to that time after you were born when you were still operating as an empathic and vibrationally sensitive organism, before you learned words and built up layers of explanation, identity, and coping mechanisms. You were merged with your environment and, like a dolphin, sent out your own brand of sonar, learning to navigate by what expanded without limit into the world and by what bounced back to you. You were like a miniature sun, beaming out clear light, love, and joy to anyone who wanted to be in it. As your unconditional love reached your parents or other influential people around you, wherever they had maintained their ability to love and feel empathy, your sonar passed through and your reality was validated, perhaps even magnified. You could feel the truth of your soul there in that particular sensation: it might even be called pleasure. A healthy childhood is marked by this kind of soul validation.

But wherever your parents or others had learned to be afraid, shut down their hearts, developed distrust, or rejected joy, your sonar bounced back to you. Perhaps your father shut off his playfulness as a child because his own parents were stern, and he unconsciously took on that strictness and the belief that all children need discipline. When your boundless joy hits his rigid beliefs and "thickened" emotions, you can't experience yourself. Instead, you sense something is wrong but don't know enough to know what it is. When it first occurs, that bounce-back produces in you-the-organism a strange sense of disorientation and lack of reality. Here is something that is "not-self," that is slower, denser, sharper, and darker than what you are. Here is the beginning of the experience of an outside world, of separation from life, of fear and an ego. These are foreign sensations—and they feel like pain.

You-the-organism are energy efficient and looking for love to sustain you. You learn that your energy doesn't bounce back uncomfortably if you adapt your behavior to match your parents' beliefs and unconscious body postures. You copy them and stop trying to express yourself when you can't get through. You won't be expansively creative if you're punished for it. You stop being affectionate if it makes your parents uncomfortable and rigid. You stop radiating warmly from your chest or eyes if your mother's eyes are unresponsive or your father's heart is hard. You learn to be silent because your mother is more relaxed then, or walk like your father because it validates him, or act funny because the moments of laughter feel better than the absences created by your workaholic parents.

You learn to flow your energy where it can flow and resign yourself to no movement in many areas. You mistake agreement with your parents' beliefs, emotional biases, and body postures for love. If you can't get soul validation, you take what you can, even if it's only crumbs, to survive. If you encounter out-and-out abuse, the intense contraction responses in your organism can severely stunt your growth, physically and spiritually. In this way, your early feeling habits begin. You hold your body a certain way, run only a certain kind of energy and intensity, and allow only a percentage of your full self to radiate without monitoring it. You define yourself as *this* kind of person and life as functioning according to *these* rules.

We are kept from the experience of Spirit because our inner world is cluttered with past traumas . . . As we begin to clear away this clutter, the energy of divine light and love begins to flow through our beings.

Father Thomas Keating

Now It's Time to Undo the Damage

The more you unconsciously adapted yourself to living within limited circumstances, with conditional love and partial acceptance, the more likely you are now to feel overwhelmed by the new, faster vibrational stimulation. These new energies seem foreign. Stimuli that don't agree with your original sonar blueprint are experienced as wrong, threatening, self-negating, and mind numbing—the same way you experienced the first sonar bounce-backs when you were a baby. As you're flooded with today's fast frequencies, somewhere deep inside your subconscious is saying, "Huh? This does not compute!" and is experiencing either confoundment and consternation or self-preserving rejection of the incoming frequencies.

Remember how the early stages of the transformation process progress? As the energy in your body accelerates, your low-frequency subconscious blocks—which tie closely to early compromises, resignations, and feelings of loss of love—can't remain suppressed. They surface and become conscious. Now you can review the outmoded decisions and misperceptions and change them. Accommodating faster frequencies, adjusting your ideas about how much love you can be and share, and dissolving your contracted feeling habits are the main work now. Yet unlearning old habits and developing new ones are not overnight propositions—they take patience, repetition, and compassion. Are you up for it? You're really going back to the openness you knew as a baby and reinstating your soul as your rightful self. Now you're going to consciously know who you are and why you're here.

As you identify your fear-based feeling habits, you'll notice that in some cases they originate with a phobic flee-and-avoid decision, in others with a counterphobic fight-and-control decision.

Unhealthy Feeling Habits Based on the "Flight" Decision

Here are some unhealthy feeling habits based on a subconscious flee-and-avoid decision:

- You literally leave the room, your relationship, your job, or the country, or drop out of the conversation, your body, and your life. You abandon yourself and others. At worst, your personality becomes split and dissociated. You feel disoriented, lost, depressed, unmotivated, and apathetic. You have memory lapses and loss.
- You distract and numb yourself with addictions to alcohol, drugs, food, sex, exercise, television, work, shopping, worry, socializing, or the Internet, to name a few.
- You live in other realities than your own or glamorize other places and time periods, celebrities, heroes, nonphysical realms and beings, or your past lives.
- In a related behavior, you're overly sensitive to displeasing the people you've glorified, so you develop codependent relationships in which

you virtually live in the other person's life, merge with them, lose yourself, and end up feeling dominated.

- You get involved in a larger-than-life situation that monopolizes your
  mind and offers a great excuse for not having to feel something: you're
  sick, in pain, incapacitated, depressed, shy, bankrupt, injured, in an abusive relationship, or have to take care of an ailing relative, a flailing
  business, or a problematic house you're remodeling.

- You feel helpless and victimized, with no personal boundaries, sense of
  self, preferences, or freedom, and gradually become drained, exhausted,
  and paralyzed. You complain, feel unlucky, make excuses, are often in a
  bad mood, and are plagued by repeating negative cycles.

- You're overtaken by anxiety and panic attacks that in time wear down
  your physical health.

Whenever you are feeling an extreme emotion, whether it be anger, hurt,
despair, or even bliss, and you act on that feeling, there can be no clarity.
The feelings are here to be felt. True action or inaction has to come from
what lies beneath all feelings. . . . In the core of the matter, in the core of
any emotion, however horrendous, there is peace.

Gangaji

Unhealthy Feeling Habits Based
on the "Fight" Decision

Here are some unhealthy feeling habits based on a subconscious fight-
and-control decision:

- You project blame, anger, rage, hatred, and violence outside of yourself
  onto other people when you don't want to feel something. You attack
  people who threaten the way you want your reality to work.

- You become a problem solver and try to fix the things that bother you
  in order to make the outside world align to your preferences.

- You launch into action as a rescuer, savior, or healer, projecting *your* ideas
  of how happy and healthy others should be onto *their* growth process.

- You try to influence or control other people through charm, seduction,
  false humility, trickery, negotiation, manipulation, or force.

- You try to know more and do more, reading voraciously, taking seminars, working long hours, or volunteering for extracurricular and charitable projects.
- You become trapped in conflict and polarized positions that don't resolve. You argue, criticize, fight, and want to reject or punish others.
- You talk yourself into liking self-sacrificial realities you think you *have* to live with. In fact, to feel in control, you may tell others you chose them or created them intentionally.
- You become stubborn and resistant, stoic and immovable. You won't change, won't listen, won't participate.

In Contrast, Here Are Some Healthy Feeling Habits

If you've remained open or are actively developing your ability to sense the world consciously, you may recognize—or want to develop—some of these healthy feeling habits:

- You allow yourself to feel, think, and act without self-judgment or self-sacrifice in accordance with a natural inner sense of ethics and harmony.
- You allow others to feel, think, and act as they do without needing to sacrifice their authenticity for you.
- You consciously notice the frequencies of other people, places, and situations and don't change yourself to match them, especially if they are lower than yours.
- You merge with other frequencies (in people, places, or situations) intentionally to activate new knowledge and ideas, then come back to yourself and determine how, or if, you want to implement the information.
- You keep your awareness open and in the present moment, receptive to subtle signals that come from your body. By remaining centered and alert, you have no need to separate yourself from the world or defend yourself from the world; you simply adjust your frequency to the vibration that most nourishes or entertains you in any given moment.
- You withdraw inside yourself, not to escape feeling something, but to see what feels right for you in each situation and to find your own ideas. You find the insights and messages in various feelings, seeing every sensation and perception as useful.

- You trust and enjoy yourself so there's no need to impress others; the resulting ease facilitates increased clarity. You trust you'll know what you need to know just when you need to know it. You share energy and awareness easily with others.
- You allow yourself to experience the full range of sensations, from the most contracted to the most expanded, knowing that this is what life is all about. Even contractions contain useful information and energy.
- You know that life and the mind naturally oscillate, so you allow yourself and others to have polarities and change often while penetrating to the core to find stability.
- You know your experience is up to you and the choices you make. You know that by adjusting the frequencies you run in your body, emotions, and thoughts, you can change your reality and create or dissolve the "stuff of your life."
- You're excited about having a live connection with the world, and you engage fully with discovery and creativity. You accept that you and the world are made of many dimensions and frequencies of awareness and that you have access to all of them.

How to Turn Unhealthy Feeling Habits into Conscious Sensitivity

By contrasting your unhealthy feeling habits with healthy ones, you'll be able to set some sensitivity goals. First you'll want to identify what isn't working, understand compassionately why you got stuck in those dysfunctional patterns so you can forgive yourself and others, and then identify some new and better methods that can replace the old patterns. The next step is to practice catching yourself when you're in the middle of an unhealthy pattern—without using another unhealthy feeling habit to deal with yourself. If you walked out on a friend because she didn't agree with you, don't go have a beer to calm down. Try a healthy feeling habit instead, like recentering and feeling into your friend's reality to seek understanding.

Don't criticize or punish yourself! Instead, say, "Oops, there I go again, isolating myself from the crazy general public," or "I notice that I want to blame my partner because she wasn't alert enough to what I wanted." Just bring it to your attention, just notice and be with it for a moment. Learn to suspend the unhealthy feeling habit, let your mind get soft like a limp muscle, and experience fully whatever your body is experiencing without trying to change it. If you feel uncomfortable, feel uncomfortable for thirty seconds longer than you previously would have. Be an observer. Bring your unconscious habits into consciousness where you can do something about them.

Anxiety is the experience of growth itself . . . Anxiety that is denied
makes us ill; anxiety that is fully confronted and fully lived through converts
itself into joy, security, strength, centeredness, and character.
The practical formula: Go where the pain is.

Peter Koestenbaum

By paying attention and "being with" the somewhat embarrassing experience of controlling or avoiding, you give yourself an opening, a slight pause. In that breather, you have the opportunity to choose an alternative. Is there a more positive way to sense what's really going on? Is there a healthy feeling habit you'd like to substitute? Instead of exploding at the person whose interests differed from yours, how about writing in your journal about what childhood experiences connect to this one with a common thread? How about feeling your body tensions and releasing them through yoga or tai chi? Once you've succeeded at eliminating an unhealthy habit, make sure you consciously substitute a healthy one. You might even want to tell yourself aloud: "See, Self? Now I'm doing it THIS way!"

Try This!
Practice a Healthy Feeling Habit

- The next time you find yourself reacting with a flight-or-fight response to intense energy or situations that are symbolic of past wounding, notice what the unhealthy feeling habit is. Suspend the action in midstream. Describe to yourself what you feel: "I'm anxious because I'm going to be late, and if the people think badly of me, it could damage my chance for success with them."
- Imagine how you're going to feel and what might happen if you continue the unhealthy habit. In this case, you might get so wound up that

you have a panic attack, get lost, have an accident, or get so frazzled on your way to the meeting that you make an even worse impression and alienate everyone. Choose to change your strategy.

- Breathe! Come back to the present moment and your body, and relax.
- Think back over the list of healthy feeling habits above and pick one. Turn it into an affirmation; for example, "I know my experience is up to me and the choices I make." Just focus on that, attune your sensitivity to how it feels to be that way, and see how your experience of the current situation changes when you apply it.
- At another time, pick one of the healthy feeling habit ideas from the previous list and practice it for a whole day. Again, turn it into an affirmation: "I allow others to feel, think, and act as they do; they don't need to sacrifice their authenticity for me." How does your sensitivity improve?

The Benefits of Being Consciously Sensitive to Vibrations

- Your intuitive ability and clarity are high. You make good decisions.
- Your creative and innovative abilities are high. You can easily materialize what you need and use what you're given.
- You cooperate easily and draw the best out of others. You have deep insight into others' motivations and sources of pain and what they need in order to grow and heal. You can offer people the help, advice, or understanding they need.
- Your capacity to feel into the inner dynamics of anything brings data and insight that give you an advantage in achieving happiness and success in your relationships, personal life, and business.
- You can dissolve feelings of separation and isolation, and understand how connected and mutually sourced everything in life really is. This helps you experience the soul in things and become more spiritually enlightened.

Developing Conscious Sensitivity Doesn't Take a Crisis!

Daniel had been a bank president for most of his career, and his lifestyle was full of luxuries. He was quick and smart, a great problem solver, a marketing genius, and a good leader who didn't abuse his power. When he retired, everyone was sorry to see him go. For a while, he traveled and took positions on various boards, but he knew there was more to life. He became restless and jumpy. He dove into a study of spirituality and metaphysics with his usual aplomb, treating it as a problem to be solved. He visited power spots around the world, ticking them off his list. He studied with good teachers, read hundreds of books, and practiced yoga. Then he had a heart attack. As he recovered, he wondered how this could possibly have happened to him, since his mind had supposedly been "in the right place" and he had been "doing all the right things."

We did a number of sessions together to find the root of the problem, and one of the first things that surfaced was that Daniel was applying mental force and his problem-solving skills to issues that could not be discovered or experienced that way. The more he used his strong will and mind, the more the experience he wanted eluded him. The thought of the void he was facing after an illustrious career was terrifying. He couldn't apply excellence as he had known it and reach his goal; in fact, he couldn't even understand what his goal was because it was a state of being and defied definition. The answer was right in his face; his heart had been flashing like a red emergency beacon. "Hey!" it screamed, "Slow down. Pay attention to me. I'm the answer! Feel me. Feel what I feel. Just BE for a while."

Knowing from his heart and experiencing "beingness" were unfamiliar experiences for Daniel, ones he had previously equated with emotional weakness and bad decision making. As we worked on learning to pause his enterprising mind and help him drop into his body and the moment, Daniel realized that he had developed several unhealthy feeling habits that had kept him fairly unconscious and insensitive about whole realms of potential experience. He had learned to be a high achiever and a good athlete and to get good grades as a child to please his father, who didn't want to feel weakness in himself or anyone near him. Daniel had done so well at this that he'd locked it in as his main identity. He had stopped radiating his heart energy because his parents didn't recognize it as real, and it bounced back to him.

He hadn't realized how much his heart had been aching from being invalidated and contracted. The heart attack was really a big release of energy—in effect, it was his soul saying, "It's time to correct this debilitating situation and move on into your full self-expression. There's nothing wrong with your mind, but it must now serve the wise dictates of your heart."

When Daniel had felt too much in his days as a bank president, it had been suppressed and dismissed, and he had distracted himself with an addiction to work, wine, cars, and travel. He had left his body, except for enjoying adrenaline rushes, and lived in his head; so some of Daniel's unhealthy feeling habits came from a flight decision to avoid the information coming from his sensitivity. The unhealthy feeling habits that came from a fight decision were: (1) pretending he *liked* the reality he'd resigned himself to having, even though deep down he missed the warmth, intuition, art, and spirituality that he had earlier shut off, and (2) he made himself invulnerable by being a proactive leader, problem solver, and expert.

In time, Daniel learned to bring his attention into his body, center in his heart, and notice what he was sensing in each moment. His choices came from those insights; he stopped defining himself and allowed other people and life to have a greater impact on him, especially emotionally. In this way, he turned around a lifelong pattern and opened himself to an expanded reality. Several years later, he became so experienced in connecting with higher spiritual realms and nonphysical beings that he began acting as a spiritual counselor and guide for people who needed the same sort of transformation that he'd experienced. Knowing him before, who would ever have thought he had this new capacity in him? All his excellence now goes into service to the heart through empathy and compassion for other people. It is a natural rebalancing and an outgrowth of who he really is.

> Turning points announce themselves through a variety of
> vague symptoms: deep restlessness, a yearning with no name,
> inexplicable boredom, the feeling of being stuck.
>
> Gloria Karpinski

Developing healthy feeling habits and conscious sensitivity doesn't have to take a crisis—unless you're extremely resistant to change. If you begin now to notice and trust more of your subtle perceptions and see how they connect to an improved, more fulfilling life, you'll gradually open like a flower blooming in the new warmth of spring.

It Helps to Find the Hidden Misperceptions

Under every unhealthy feeling habit is a misperception about the way energy and awareness really work. Remember that when the habit was originally established, you didn't even understand language and had no mental concepts; everything was visceral, instinctive, and survival oriented. Your reptile brain and animal nature were keeping you alive. There could have been no other way to get through childhood. It's absolutely normal that as you reach maturity, you reconsider these early patterns and clear them so you can become the being of light and love that you really are. Now is the time when you can feel into each habit to find the underlying misperception made by your baby mind. Once you connect compassionately to the odd logic and the reason you formed the habit, it's much easier to let the habit fade and even chuckle about it.

For example, at the beginning of this chapter, I mentioned Megan, whose father left her when she was small. Without being able to understand the reasons behind his actions, her body contracted in a high level of confusion and overwhelm about the loss, while her baby mind had these perceptions: "My male loved ones leave me," "I caused it because I'm not good enough," and "I can't have male energy to help me, so I'll have to do everything myself." She distracted herself from sensing these things through the unhealthy feeling habit of acting rebellious and tough, yet in spite of her efforts to avoid abandonment, her boyfriends did have a tendency to leave her, she did feel she was doing something wrong, and she did feel she was being overly independent. Finding her early misperceptions can help Megan turn the pattern around so she can have a supportive marriage and rewarding career if she wants to.

In Daniel's situation, he became the strong, competent, infallible one so he wouldn't lose his father's love. Since his father's heart was shut down, approval and agreement were the best Daniel could get in the way of love. His unhealthy feeling habits were to distract himself and be stoic. The underlying misperceptions that his baby mind made were:

"Men don't show love or tenderness," "Success is necessary for survival," and "I'm good and safe because I'm smart." If we keep going, underneath those misperceptions we find another basic misperception common to just about everyone: "This world is a place where I must sacrifice my true self and suffer because of it. The experience of being alive is one of pain."

It's interesting that as Daniel transformed, he reversed every one of his original misperceptions. Men do show love and tenderness; it is one of their great joys. Material success is not necessary for survival; being purposeful and expressing authentically are. I'm good and safe, not because I'm smart, but because I'm being myself. This world is only a place of sacrifice, suffering, and pain if I say it is.

Here are a few examples of other basic misperceptions about life:

- The world, with its insanity and cruelty, is bigger than I am, and I'm helpless against it.
- If I feel their pain, I'll feel mine. If I feel mine, I'll die from the hugeness of it.
- I need to prevent others' pain, or they'll abandon me and I'll die.
- No one wants me, so I have no meaning or worth.
- I need to remain invisible, or I'll attract punishment; when I show up, I am wounded.

Try This!
Reverse Your Basic Misperceptions

- Feel into your unhealthy feeling habits and see if you can sense the basic misperceptions your baby mind made when you were tiny. Write them as statements. You might notice how those statements have played out in your life.
- Reverse each misperception so the opposite is true. Write each as a positive statement. Don't forget to include: "This world is only a place of sacrifice, suffering, and pain if I say it is."
- Focus on experiencing each positive statement as though it is your new truth as many times a day as you can remember to do it. Notice how your urges, motivations, ideas, and actions shift.

Just to Recap . . .

You were born vibrationally sensitive and empathic; it's the way your heart and body formed and how you learned to become part of the world. Your natural state is to be sensitive, but you may be more or less aware of it. You may be an unconscious sensitive, a conscious sensitive, or a highly conscious sensitive. You may have developed unhealthy feeling habits—which distorted your sensitivity but helped you survive—when you were small and nonverbal. You developed these feeling habits in much the same way that dolphins use sonar; you radiated the qualities of your soul, and where they matched those of your parents, you could feel yourself and felt validated. Where they didn't match—because of your parents' fear—the energy bounced back to you, confused you, and created the sensation of pain. You learned to match the fear vibrations in your parents to find approval or agreement: a poor substitute for love.

You have both healthy and unhealthy feeling habits. The unhealthy ones are in the way of your transformation process because they keep you contracted in fear or in partiality. Your unhealthy feeling habits are based on a choice to either flee and avoid or fight and control. Today's increasing energy frequency speeds the surfacing of your subconscious "wounds" and brings an urgency to clearing the underlying misperceptions your baby mind made before you understood you were in charge of your own experience. You can turn the unhealthy habits into conscious sensitivity by catching yourself in a reaction, being with the experience, and choosing an alternative healthy feeling habit to experiment with instead.

Home Frequency Message

As I explain on page xxi in *To the Reader*, I've included these pieces of inspired writing at the end of each chapter as a way for you to shift from your normal, speedier reading mind to a deeper kind of direct experience. Through these messages, you can intentionally change your personal vibration.

The following message is meant to transport you into a way of knowing the world that's close to the way you'll experience life in the Intuition Age. To move into the *home frequency message*, just downshift to a slower, less hurried pace. Take a slow breath in, then out, and be as calm and still as possible. Let your mind be soft and receptive. Open your intuition and prepare to *feel into* the language. See if you can experience the deeper realities and feeling states that come alive *as you read*.

Your experience may take on greater dimension in direct proportion to the amount of attention you invest in the phrases. Focus on the words a few at a time, pause at the punctuation marks, and "be with" the intelligence delivering the message—live and right now—to you. You might speak the words aloud, or close your eyes and have someone else read them to you and see what effect they have on you.

ENDING SUFFERING IN YOURSELF

Suffering, even great physical pain and emotional loss, dissolves when you stop looking at it and for it. It disappears when you stop agreeing with it or fighting it. It becomes nothing when you cease to extract identity from it. In the higher realms, there is no suffering, there are no victims, no saviors, no leaders, no followers, no haves and have-nots, no here and not-here. In higher frequencies, you know the power of being. Being reveals presence, and presence reveals divine awareness watching over, and from within, all. Divine awareness reveals love as the basic unchanging nature of self and life. In any moment, you can Be. You can look for and feel presence; you can expect the surprise return to the love you've never left. There is no pain in love, only in separation from love, and feeling separate is the choice to suffer.

In the higher realms, your self is our self is the Self, and the Self is one shared experience of loving presence. If you think you are not the one Self, or that another is not the one Self, that it is possible to be lost, then you suffer. Thinking of "them," or a space between, or a vacuum, makes an artificial gap, because there is nothing outside you, no strangers, nothing foreign, nowhere that presence can escape to make an emptiness. As soon as you pretend to have a gap in the continuum of presence, you create suffering, which is doubt and fear, which is separation, which is insanity. It is only in the mind. The moment you refamiliarize the world, melting it back together again within yourself, the gap dissolves and indivisible presence reappears. One way you see yourself as a small, finite trembling self, the other way as an unlimited, ever-expanding radiant self.

Sometimes you will feel pain, which is the resistance to natural contraction. You need not suffer, though. Release and move on with the wave. Sometimes people who suffer will occur "within the space of you," in your world. This means you still contemplate the possibility of suffering's reality. Be with them fully, for one pure moment. Don't believe them, don't negate them, don't copy them—just allow their experience. Feel the presence in them, let them feel your presence, offer it as the presence of the one Self, and let them merge into your offering as much as they wish. As presence becomes conscious, calm certainty emerges, and they remember their true self. They find their way out of suffering, and this frees you, too. As the memory of love occurs, there is instantaneous healing, sudden positive change, and personal transformation—in them and you. Every dissolution of the illusion of no-love erases suffering for all of us. Every collapse of the gaps the mind makes, and suffering dissolves, like a cloud into the sky's deep blue.

4

Freeing Yourself from Negative Vibrations

The full and joyful acceptance of the worst in oneself may be the only sure way of transforming it.

Henry Miller

Claudia walked into the meeting with a prospective new client and immediately felt clammy and cold. "I'm picking up some bad vibes," she thought. She pushed the sensations to the back of her mind and readied her presentation. As she began to speak, she noticed that one member of the prospective client's team was leaning toward his buddy, whispering something with a snide expression on his face. The second man glanced at her blankly, then returned his gaze to his electronic calendar. Suddenly Claudia felt judged and ill at ease. Self-conscious now, she wondered what they might think was wrong with her. Wasn't she attractive enough? She pushed her anxiety down with an act of will and gave the most intelligent, charming presentation she could muster. Looking around the conference table afterward, she watched her own teammates take over, acting like they had the inside track, adding information she hadn't been aware existed. She was being upstaged. Now Claudia had a knot in the pit of her stomach. She felt she'd been tossed into a tank of sharks.

For the rest of the day, she felt left out and couldn't concentrate. That night she couldn't sleep as her mind imagined criticisms and created negative scenarios. "I should have worn my red suit." "I need to be funnier." "Something terrible is going to happen to me." The next day, just as she suspected, she was called into her boss's office and dismissed; he said she wasn't performing at a high enough level and didn't get along well with her colleagues. As she left, Claudia thought the previous day's meeting had certainly been a setup, designed to belittle her so they could fire her. Her mind spun out of control, reviewing times in the past when she'd been used and disrespected, conjuring up a limited future in which finding a new job would be difficult, she'd have to take a demeaning position, and she'd end up a disillusioned old woman like her mother. Then she shifted into a fury aimed at her colleagues, who could so easily manipulate circumstances toward their own ends.

After days and weeks of this negativity, and a variety of unsuccessful job interviews, Claudia got sick with a lingering flu. Now she couldn't job-hunt, and as time passed, her finances dwindled, scaring her even more. She couldn't see a counselor or even get a massage because she didn't have the money. She wasn't making a good impression when she looked for work because she was giving off such strong "neediness" energy that no one wanted to be around her. This added to her depression, and she spiraled down. Claudia was stuck—so stuck that her mind could see no way out. She repeatedly ran the tape loop of how stuck she was, how she didn't know what to do, and dug herself in deeper. She had to *do* something! But everything she tried failed. But she had to *do* something! But she'd tried everything already. But she had to *do* something! But she couldn't afford to do anything. But . . .

How Do You Get Stuck? Let Us Count the Ways!

Does Claudia's downward spiral sound familiar? Have you ever painted yourself into this sort of seemingly inescapable corner? If so, you probably activated all your unhealthy feeling habits and gradually lowered your personal vibration down the scale toward paralysis, numbness, or even suicide. When your frequency winds down and seems to stop, it's hard to see how things could possibly change and difficult to remember that you are an unlimited soul with a fabulous destiny that you're *quite* capable of materializing. But how do paralyzing conditions happen? Let's examine the inner dynamics of being caught in negative vibrations, then we'll look at ways of reversing the patterns so you can more quickly extricate yourself from these periods of truly wasted time.

Have you learned lessons only of those who admired you,
and were tender with you, and stood aside for you? Have you not
learned the great lessons of those who . . . braced themselves
against you . . . or disputed the passage with you?

Walt Whitman

Any number of seemingly innocuous experiences can shock or scare you at a deep level without your realizing it—and this can cause you to recoil and contract. The two men who judged Claudia without speaking a word frightened her at a subliminal level, and she went from a high vibration of eagerness and enthusiasm to a contracted, lower vibration of self-doubt in a matter of seconds. When you have negative experiences, there's a natural tendency for your personal vibration to drop and for you to feel "bad." And when your frequency is low, you tend to have more negative experiences, which leads to an even more negative personal vibration, as witnessed by Claudia's slide from disappointment and frustration into hostility and desperation.

Add to this the mind's penchant for labeling and locking in the experiences as "bad," "wrong," or "painful," then associating your identity with those labels. Now you are "bad," "wrong," or "in pain." That's no fun, so you use your willpower to resist or escape the experience you're having, and the flow of your life grinds to a halt. A low personal vibration blocks your soul's reality, and you suffer from lack of wave movement. Here are seven ways you might get stuck.

1. A cycle of growth is complete but you haven't realized it yet, and you haven't let go. You've finished creating something, developed new skills, learned some lessons, and your vibration is shifting to a new level—but like Claudia, you haven't realized that you're finished and that you are the one pulling out of this chapter of your life. Maybe you're too loyal, or haven't realized you're bored, or haven't used your imagination to flesh out a better reality. If you haven't gotten the message that the job is over, life will end it for you and you'll be fired, like Claudia, or be forced to quit because your mother gets sick or your husband's company transfers him. If you don't recognize the trough, or low point of the wave, and choose to go along with it, it's easy to feel that life is rejecting you or that you've failed. If you use endings to criticize yourself, feel unloved, or fall headlong into fear of the void, you'll spiral down like Claudia did. Every natural cycle of creation involves materializing an idea into form, then dematerializing it to create a clean, fresh space or new blank canvas. We forget that we create form, then space, then form, then space. We forget how necessary and pleasurable space is.

2. You don't use what you've been given, or you want more than you can use. It's common to be out of synch with your natural growth cycle, protesting that what you've received isn't what you wanted, trying to force something to happen before it's time, or keeping something in a form that wants to dissolve. What you have now is what you've been consciously or unconsciously thinking about and focusing on in recent days, weeks, and months. By protesting and complaining about your current situation, you invalidate the perfect functioning of the materialization process, which "stops" your *life-wave*, or your natural flow.

Your life emerges out of your personal vibration, so if you're plagued with troublesome experiences, you may have been steeped in negative emotion and thinking for a while, and the snagged experiences exist to remind you of this. You probably need to raise your vibration by focusing on things you love and the way you prefer to feel. Ask yourself: What habitual thoughts or feeling habits have brought these results? What am I showing myself? By using what you've created, you can consciously complete a creation cycle and open space for the new to emerge. You can also block your life's flow and get stuck by wanting more than you need to learn your next life lesson. Creating too much means you'll just have to get rid of it, and that wastes time and energy.

3. An experience triggers a memory of an earlier similar negative experience. When life triggers a painful memory, you leave the flow you're in and "teleport" back to the painful past. Your subconscious reacts automatically with an unhealthy feeling habit that mimics the way you dealt with

the original experience. You refuse to feel what's surfacing by indulging in fight-or-flight behaviors where you don't trust yourself or others, can't sense key guidance, and/or feel overwhelmed, hyperactive, defensive, distracted, disoriented, depressed, or victimized. You can paralyze yourself with blaming, punishing, controlling, or being stubborn. In Claudia's case, her dismissal triggered memories of her mother—whose alcoholism ruined a promising singing career—and Claudia became possessed by the unrealistic fear that something similar might happen in her own life.

4. You talk about what's not, what hasn't happened yet, what might never be, what you don't like, what you don't do, or who you aren't. When your mind focuses on describing nonexistent or empty realities, you are, in effect, materializing *nothing*. Plus, you've abandoned your life's flow; it can't move without your conscious presence. To experience something as real, and to therefore have motivation and action, your conscious mind must be centered in your body and perceiving through your body, which is your reality filter. When your self-talk describes a negative or nonexistent reality, your body can't grasp it because bodies *do exist* in time and space. Your body gets ready to act on a brilliant idea, for example, until you say "I'm *not* smart." The same thing happens when you say, "I want to make more money, *but* I don't know how." Then your body consciousness goes into the same kind of consternation it experienced when you were a baby and your love bounced back to you unmet. Huh? It cannot comprehend the concept of negativity and nonexistence, and as it struggles to understand what it's supposed to do, you get stuck.

When you talk about emptiness and nonexistence, you create an imaginary hole or gap in yourself and your world. That gap is a place where you pretend not to know or experience your soul. Lack of confidence is such a gap, as is pretending ignorance, fearing loss or difficulty, or avoiding direct experience that involves body, emotion, mind, and soul. Gaps give you the illusion of impossibility, which discourages self-expression and stalls the movement of your life-wave.

5. You project your mind and energy into other people's lives, other places and times, or fictitious realities. Similar to talking about what isn't, projecting away from the here and now into other times and places also removes you from your body and your own process. Your conscious mind is elsewhere and cannot receive insight, instruction, and love from your soul—which flows to you via your body—because you are not "in" your body. Since "nobody's home" to receive higher guidance, your subconscious mind will take over and automatically react to situations as it did in the past. You become reactive instead of responsive, and this calls up many of your unhealthy feeling habits, as previously mentioned. What worked in the past doesn't necessarily work for today's situation. In addition, knowing what's needed in someone else's life doesn't mean that the same thing is good for you. By not living in your own body and life, you postpone your growth and end up feeling stuck.

6. **You're attached to your creations, habits, definitions, and identity and are in some sort of "holding pattern."** You may be *holding on* to beliefs or possessions for security, or *holding back* from expressing your truth or love, or *holding forth*, controlling the people around you by expounding on a subject. Maybe you're *holding to* a vow or commitment you made in the past that is no longer relevant, or holding on to a resentment, grudge, or injustice, or holding back from changing. You may be holding your breath, hoping something will or won't happen, or holding yourself in check so you don't express yourself too blatantly. Whenever you contract your energy and try to pause or stop a wave, you slow your own flow and lower your frequency, which causes you to miss opportunities. You also wear yourself out damming up the river. When holding patterns become severe they turn into obsessions and fixations.

7. **You are drained from too much negativity, conflict, and willfulness.** Perhaps you have so much stress that you can't sleep. Maybe you're involved in a heavy addiction or plagued by chronic pain. You might be fighting to maintain yourself against bullies or trying to prevent a failure. The more dissipated you become, the less motivated you are and the less energy you have for positive imagination. When your awareness is dull, your emotions become apathetic and your body becomes lethargic. You've painted yourself into a corner and temporarily feel incapable of movement.

Getting Unstuck Has Never Been Easier

In spite of struggles and difficulties, please remember that you're floating down a river headed for a destination that is a sure thing: knowing yourself as your soul. You are in the part of the transformation process where old low-frequency fears and unhealthy feeling habits surface and want to move out. You are experiencing the reaction to this, in which you want to resuppress the discomfort and pain through fight-or-flight methods. Your struggle against suffering and fear of the void wears you down, however, and eventually you enter the phase where you clearly see that the old habits and forms have outlived their usefulness.

> **The old world is slow because of separation, gaps, and fear.**
> **The new world is fast because of interconnection.**
> **As we approach unity, life is more instantaneous.**
> **Our lives work by new rules based on the**
> **speed of the present moment.**

Many parts of your life and ideology dissolve. You're more willing to let go and see what happens because you're tired. As you release struggle and find peace, you glimpse your new reality, which motivates you to keep on. Yet you may still trip every few steps, so you need to know you can free yourself easily from negativity whenever you notice you're caught in it. Here are two reassuring ideas.

First, since your personal vibration is evolving to a higher frequency along with the planet and everyone else, you will simply not be able to stay stuck as long as you used to. When you feel stuck, or fear an ending, remember that you exist within a life-wave, that life is always moving. Waves have high crests and low troughs, and stuckness is just an indication of an impending turn in your wave. Nothing ever truly stops; you can count on an ongoing momentum to move you repeatedly and ever more quickly now, through ups and downs and beginnings and endings. The more you allow yourself to relax into the current and feel that everyone and everything is somehow in the same flow with you, the easier the turns will be.

Second, because the accelerating process is working with you, clearing fears and subconscious blocks happens much faster than it used to. We used to have to go into therapy, talk about our problems for years, and have cathartic emotional releases. Now the therapeutic process can occur much more quickly. By shifting into the present moment, to a fresh space where you let go of your personal history of woundedness and choose the soul's reality instead, healing is seen as a repeated choice for soul and the release of a false reality. Healing or "recovery" is in the present moment instead of the future; thus, it doesn't require such a long linear process to return to your best self. Before you know it, your story of suffering loses its meaning and disappears. We'll be delving more deeply into this centering process in chapter 5. Meanwhile, there are a number of ways to remain in the flow, free of negativity.

> When we quit thinking primarily about ourselves and our own self-preservation, we undergo a truly heroic transformation of consciousness.
>
> Joseph Campbell

Free Yourself by Cooperating with the Waves

Since waves go where they want in spite of obstacles, it's best not to get in their way. In fact, it's more useful to think of yourself as the wave itself, rather than as the immovable object in the wave's path. You are a vibrational being moving through a sea of vibrations. You must master the art of blending and aligning with the constantly changing dynamics of living energy. And that means you need to learn a totally new way to use your willpower, because it is willfulness that so often makes us act like brick walls.

Learn the right use of willpower. If you're like me, you probably internalized the voices of a variety of authority figures in your youth, who now remind you about what you "should" do. I particularly remember being told, "God helps those who help themselves," and how it took me a while to get my head around the concept, as it seemed that if I helped myself I wouldn't need God. But at any rate, this notion set me off on a path of self-sufficiency where I relied heavily on an iron will to get things done. I used my will to project and maintain my intentions, to be diligent and disciplined, to not give up when the going got tough. "Where there's a will, there's a way"—that was another homily that stuck with me. After years of maintaining my success through focused and tightly held willpower, I was exhausted. I didn't want to do *anything* anymore. I was sick of all the "should" voices. It was then that I realized there is a "right use of will."

Shifting out of negativity and stuckness is often a matter of not exerting the old forceful willpower but "choosing what's choosing you."

Willpower is not about resisting, forcing, or controlling—it's about choosing. And there are just two basic choices: to feel expansive, loving, and connected to the high vibrations of your soul—to literally *be* your soul—or to feel contracted, afraid, and immersed in the low vibrations of suffering—to *not be* your soul. If you choose to feel alone and separate, you'll assume you must do everything under your own steam by controlling yourself and the world, as I did for many years. If you choose to feel connected to life, you'll need very little of the old type of willpower. You'll discover that concepts like flow and synchronicity take the place of willpower.

When you're stuck, you don't have to willfully figure out a whole strategy for how to change things and what the new phase will look like—all you have to do is suspend the old, contracted, low-frequency pattern. Use your willpower to choose to be with what is. Imagine that your mind is a muscle and let it go limp. Say "Duhhhhhhh" and let your mouth hang open! Oddly, I've found this helps blank out the over-busy mind that thinks controlling things is the answer. In the quiet space of allowing, you can easily notice what the wave wants to do. Are you at an end you haven't recognized yet? Is there a new inspiration knocking on the door that wants to become conscious and take shape?

Follow the flow; it knows where to go. We are rocking into form, then back into energy, and back to a new form, then back to energy. Every other millisecond, every other minute, every other day—whatever amount of time you want to focus on—you have a new chance. Life's waves help you let go and open new space, help you find new fascinations and motivations and get started again. All you have to do is trust the wisdom and plan behind the energy flow, then go with the river in the direction it's heading. Sometimes it rages with rapids and whitewater, other times it disappears underground, then it reappears and drifts along peacefully. Sometimes you're being, sometimes doing, sometimes having.

Life is a process of becoming, a combination of states we have to go through. Where people fail is that they wish to elect a state and remain in it. This is a kind of death.

Anaïs Nin

Recognize crests and troughs (beginnings and endings). I've learned after torturing myself—and finally deciding I don't like being tortured—that as a self-employed person, my workflow parallels my inner needs. I used to think life was capriciously "having its way with me," but eventually I saw there was another principle operating. If I'd been in a period of intensive work with clients, for example, and was beginning to complain that the crammed schedule was a problem, the truth was that I was ready to shift into a spacious phase with some alone time during which I might process insights, collect myself, and gestate new projects. Then my phone wouldn't ring, or work I had tried to get would fall through, and I'd have my quiet time. I usually wouldn't recognize it as a personal need; I'd feel I was being punished by having this "slump" and would make the gift of my alone time into a problem. "What's wrong with me? Don't people like my work? I *have* to make more money!" I'd mutter. If I stayed in complaining mode, the period of "space" would be prolonged. If, instead, I thanked myself for the renewal time, saw what I'd learned, and realized I was now hungry for people, communication, and external stimulation, the cycle would shift easily and opportunities would show up on my doorstep within days—sometimes hours.

Moving harmoniously with a wave involves mastering the turns, both at the crest and the trough, seeing that you're about to receive the next thing you need rather than thinking you're moving from one problem (like overactivity) to another problem (like underactivity). The turns are where you can realize the gifts that have been given and the lessons learned, so gratitude and optimism are particularly useful at these times. The turning point at the crest is when you reach the most materialistic, extroverted view of life, when materialization is complete and you feel "high" and successful. In physics terms, the wave has become the particle.

The most challenging time may be when the wave turns at the trough—when you're bored, feel things fading, need space, and must release meaning and what's outmoded in order to return to Being. In studying physics, this is where the particle becomes the wave. Moving from trough to crest seems like the fun part because it involves enthusiasm, motivation, and achieving goals. But releasing old forms, relaxing, dreaming of multiple imaginary realities, and rejuvenating ourselves are every bit as pleasurable. Chronic resistance to the turning points of a wave can cause exaggerated dramatic shifts, such as crises and traumas.

> For emotional healing to prevail, we must enter fully into whatever
> is left unprocessed and ride it all the way through until the pent-up
> energy is drained off the issue. Then, it is finished, never to
> command your attention or psychic energy again.
>
> Jacquelyne Small

Encourage fluidity wherever you find stagnation and frozen awareness. Watch for holding patterns. If you're withholding your ideas, try starting more conversations and sharing more with others. If you're too sedentary, get on that treadmill! If you're holding on to an old lover because you don't want to be alone, "get a life" and develop new interests. If your aging father is fixated in a habit of watching television all day and complaining about having no friends, take him to the YMCA for a swim. If your family is in a rut, vary your routines and switch responsibilities, or rearrange the furniture. If you're feeling too analytical or verbal, try shifting to a different part of your brain: go for a "reptile brain walk," letting your body decide whether to turn left or run for no reason. Move into your senses—try smelling things for an hour.

**Forcing something or "holding" is an indication that you're out
of harmony with the flow and missing some key information.**

Watch your mental habits, too. Do you make a lot of declarative statements and pronouncements? Do you categorize people based on first impressions? Do you label your experience and compulsively lock down meanings too quickly, saying "This is just like that"? Whenever you define and label something, it stops moving and has less chance to evolve creatively. If you say "I'm angry," you lose the nuances concerning your more subtle sensitivity and body awareness. Instead, stay fluid and describe your experience: "My stomach is tight; I feel frustrated that my husband didn't listen to me. Now the sensation is making my throat tight, and I feel like I'm not allowed to express my own needs and ideas, and that scares me, and I actually want to cry." See if you can live without so much definition, or let your definitions be more fluid and temporary. See if you can experience life directly. On the other hand, if you habitually say, "I don't know," you might look at how this also stops the wave of your self-expression and growth.

Try This!
Where Are You Holding?

- List the different ways you are holding on, what you're holding on to, and why.
- List the different ways you're holding still, holding out, holding back from doing something, what you're avoiding, what you really want, and why.
- List the different ways you hold forth or try to control your world and others—by being the expert, the center of attention, or talking too much—and why.
- List the different ways you are holding to an old vow, commitment, or rule that may no longer be valid and why.
- List the different ways you've labeled your experience, and how, if you unlabeled it, you might discover new possibilities.

Let the waves pass through you. Waves continually move through you and the field of energy and consciousness of which you are a part. There are waves that bring new frequencies of energy and information. The same waves move energy and information away as they pass through you. Event waves often reach you before the event, like lightning before thunder. For instance, you may be driving on the freeway and notice the cars around you acting erratically or becoming sluggish; up ahead you find there's been an accident. The ripples from the accident are radiating in all directions, and as people begin to feel the disturbance, they act disturbed. I remember years ago when Mt. St. Helens erupted and I was living in Northern California—for at least a week beforehand, I was abnormally irritable and angry, always on the verge of "blowing my top." As soon as the volcano erupted, I calmed down.

It's common to unconsciously try stopping a wave as it passes through you so you can look at it and see what it's all about. If an event wave is bringing you information about a volcano or an accident, you're likely to respond energetically with nervous agitation. Your mind can't interpret the energetic information and often erroneously thinks the disturbance is about you. I think I am unusually angry, when really it's a *volcano* that's about to blow *its* top! I think I'm going to die, when it's really a friend who is dying. I'm unreasonably sad when it's really the victims and survivors of a coming terrorist attack who are about to be devastated. You don't need to stop a wave to understand the information encoded in it; if you let it pass through, it will download into you as it moves. All you need to do is ask yourself regularly, "I notice I'm feeling (sad, disturbed, unreasonably happy, and so on) for no reason. Why am I noticing this?" This way, you won't have the needless distortion of thinking something is wrong with *you.*

Equalize the emphasis between the physical, emotional, mental, and spiritual areas of your life. Imagine a wave of creativity coming from the high frequency of your soul, bringing a new pattern for your life, descending through the octaves of your awareness: from the spiritual, down through the mental to the emotional, into the physical realm, and then back out again. If these levels of your awareness are equally developed, the wave passes through evenly with a constant rhythm. It downloads its content into you effortlessly. You experience good luck, easy flow, and harmonious life balance. But if, for example, you have avoided your emotional reality in favor of a life of logic, rules, and abstract concepts, your mental level will be "wider" since it's been overemphasized. The emotional level will seem "narrower" because it's been underemphasized. The wave will have to adjust its frequency to the broader, overdeveloped band of mental awareness, then readjust to the narrower, underdeveloped emotional band, then perhaps widen again at the physical level. The movement becomes jerky and dissonant, and your life process will reflect this with various snags: you might experience difficulty completing projects, finding motivation, or you might feel undue pressure or emotional paralysis. When you are evenly developed in body, emotion, mind, and spirit, the waves flow through you rhythmically and you receive the greatest benefit.

Try This!
What Areas of Your Life Need Equalizing?

1. Imagine a thermometer with markings from 1 to 100 along the side. Ask your inner self to give you a reading for how much you're using each of the following parts of your awareness, with 100 being the most.
2. How developed are you, and how much are you actively using your spiritual awareness? Your mental awareness? Your awareness of feeling and subtle sensitivity? Your physical instinct and body awareness?
3. After you see the relative percentages, list three to five ways you can increase the development and use of the underactive areas so all the parts of your awareness are even.

Raising Your Frequency Frees You from Negative Vibrations

You can get stuck in negative vibrations when you don't feel your soul's presence and your unity with life. Something is in the way—fear, lies, or misperceptions—or you experience fragmentation and empty space. Every "missing piece" of soul that you don't experience lowers your frequency. Every time you split your awareness into fragments—which means you experience separation—your vibration slows. When your personal vibration is stuck in a low frequency, you'll tend to fall back into unhealthy feeling habits, negative thinking, and sluggish physical heath.

Raising your frequency, which always frees you from negativity and stuckness, means relaxing to make time and space to experience more of your soul. You can do this just by smiling or by imagining a scale, like a thermometer, and raising your energy 10 percent. Or you can visualize a better reality, a brighter color, or an uplifting random act of kindness. You can choose to feel more accepting and generous, for example. You can experience something you've been resisting or engage with your life more deeply and mindfully. Trying to jack up and control the frequency of your personal vibration with willpower only results in hyperactivity, stress, and an eventual crash. Your frequency rises to its innately high level when you clear away your mental and emotional clutter and stop blocking it. When there is nothing in the way, the clarity and warmth of your soul shine through effortlessly.

Try This!

Raise Your Vibration by Breathing Deeply and Slowly

It's long been known that oxygen raises the frequency of your body. Also, energy moves more slowly through tight muscles, so relaxing your body and breathing deeply is key in raising your vibration. High-chest, shallow breathing is indicative of anxiety.

1. Sit upright, feeling supported, and let your muscles just hang and be at ease. Stop your internal dialogue and listen to the silence. Be still, feel subtle variations in your body, and focus on the idea that the oxygen in the air is going to supercharge your blood and make your body feel extra alive. Let your breathing be silent, very slow, continuous, and seamless—so it seems to curve without pauses at the end of the out-breath and inbreath.

2. Draw in your breath to fill all possible cavities, from your sinuses to your deep belly. When you think your lungs are full, breathe in a little bit more, filling every pocket. Imagine your rib cage stretching beyond its normal capacity.

3. Exhale by slowly collapsing your ribs, drawing in your stomach muscles, and tightening your diaphragm until you've squeezed out all the air.

4. Continue, counting from one to ten, one number on the inhale, one on the exhale. Think only of the number, and if others thoughts intrude, stop and start over. Try this for twenty minutes.

Decluttering Your Personal Energy Field
Raises Your Frequency

Your light is shining through a sieve, and you are the result of how much gets through. Think of yourself as surrounded by concentric spherical layers, like an onion. The layers very close to you contain physical information, the layers beyond contain emotional information, then comes information about your thought patterns, and even farther away are layers that contain information about your soul and life purpose. These are like octaves of your awareness. At the soul level, there is no fear or blockage— just clear, compassionate *diamond light*, a quality of light often visualized by meditators to give the feeling of pure awareness. But in the physical, emotional, and mental layers, you'll find physical dysfunctions, frozen feelings, and fixed ideas that have resulted from past experiences during which you were confused and afraid. Those contracted patterns are like shadows; they are immobile places where you don't experience your own truth and love. You'll also find holes and gaps where you're fragmented or avoiding something, and these act as blocks, too.

When you quiet your mind, you offer no thought; and when you do so, you offer no resistance; and when you have activated no resistant thought, the vibration of your Being is high and fast and pure.

Abraham/Esther Hicks

Imagine now that your soul is projecting wisdom, intention, and energy down through the octaves to create your life, body, and personality. Because of the many shadows or solid places, and the empty gaps in the field of who you are, only a certain percentage of your totality can make it through, the way light through a sieve can only pass through the openings. Everywhere there's a shadow or gap in the higher dimensions, there will be a matching contraction or unconscious place in your body and life. A memory of emotional trauma and the beliefs that developed around it will cast a shadow onto the body, perhaps causing chronic pain, illness, or injury at a location that corresponds with the original wound. For example, if a person has been repeatedly beaten, either in a past life or early in his or her current life, the memory of pain, bruising, or broken bones will remain in their more subtle "energy body," and later, the contracted energy or shadow might easily cause chronic pain of no known origin to surface in the exact areas—like the face—where the hatred and anger were so forcefully and painfully delivered previously.

When you heal by understanding and releasing the stuck emotions and beliefs, the dark spots in your field dissolve, and now more of your soul's diamond light can flow through. Here on earth, your frequency increases, you become wiser and more loving, your body heals, and your life improves. So, simply through clearing soul-blocking feelings and thoughts—your unhealthy feeling habits—your personal vibration naturally increases.

Some common soul blockers are the unhealthy feeling habits we have covered previously: being a victim or dominator, projecting blame, being stubborn and willful, rescuing others and wanting to be rescued, and avoiding reality through distractions, procrastination, and postponement. Add to these: envying others, attacking/fighting, complaining and using negative (I can't, I hate) or ugly (put-downs, gossip) language, and imagining elaborate worst-possible scenarios. The Buddhist nun Pema Chödrön calls these reactive behaviors "being hooked," like a fish taking the bait.

Every time you reverse one of these hooks or behaviors and substitute a healthy feeling habit, every time you let go of resisting and just *be* with what is, you allow more of the soul's diamond light to energize you. And every time, presence reveals important knowledge, reinforces the compassionate view, and helps you know what to do next. When you unlabel something or pull your invested energy out of a fixed idea or definition, you dissolve another shadow and more diamond light flows into your life. The same thing applies when you decide to "clean up your act" and eat healthy food, lose excess weight, and stop smoking or polluting your body with addictions.

Another category of soul blockers has to do with ideas, beliefs, and worldviews you unconsciously took on to survive in your early years. They may have nothing in common with who you really are and what you're here to do. These overlays originated back in your sonar period, where you unconsciously adapted to your parents' belief structures and body postures.

Your overlay may tell you that you have to be polite and humble when you're actually ready to become a fearless journalist. These ideas are like wet blankets weighing you down, making you act in ways you've out-grown. You have false ownership of these ideas and might imagine giving them back to the people you borrowed them from, or you see them melt away or vaporize out of your energy field. You can recognize these inher-ited ideas because they are preceded by the word "should," or when you hear yourself recite them, you hear an echo of someone else's voice.

> Your past is not your potential. In any hour you can choose to
> liberate the future . . . Ultimately we know deeply that the
> other side of every fear is freedom.
>
> Marilyn Ferguson

Try This!
Clear Away Other People's Overlays

- List the mores and values you live by, even the negative one's you justify. Which ones came from your mother? From your father? Are there any ideas that you feel are outmoded or don't really apply to you? If so, give them back to the person you obtained them from, or let them dissolve.
- List the habits and beliefs you hold about money, work, relationships, parenting, health, aging, religion, politics, and death. Where did you get these ideas? Do you need them? Try suspending them one by one. Feel how it would be to allow each area to teach you spontaneously about how to be and what to do instead of having a fixed opinion or rule. How might each area expand or change?

If you've been stuck in habits that promote ignorance, deprivation, help-lessness, absentmindedness, lack of self-worth, or complaints, there's only one thing that fills in these kinds of gaps, and that is *presence*, the quality of all-pervading lovingkindness that underlies everything. Center yourself, fill up with presence, and you'll have the "presence of mind" to counter your unhealthy feeling habits. When you hear yourself say, "I don't know," try saying instead, "What do I already know about this?" When you hear yourself telling your friend, "I'm not a good dancer," you might entertain the thought that you are capable of moving in interesting, unique, or creative ways. What would it be like to have *your version* of being a dancer as part of your life? If you hear yourself repeating your tape loop about never having enough money, you might say to yourself: "Pause! I've had enough so far to stay alive and live at a certain level. I'm fine. I can change my circumstances whenever it's interesting enough to me to do so. Am I interested in this now? What do I feel like creating?" You are the author of your own story. You've mysteriously been given the *incredible* gift of life, and at the same time, you have free will to choose your attitude, mood, and activity level. There's no force or reason in the world powerful enough to prevent you from being your full self, if you want it.

There is no rushing a river. When you go there, you go at the pace of the water and that pace ties you into a flow that is older than life on this planet. Acceptance of that pace, even for a day, changes us, reminds us of other rhythms beyond the sound of our own heartbeats.

Jeff Rennicke

Just to Recap . . .

Being stuck in negativity is caused by four things: low personal vibration, improper use of willpower, not working harmoniously with waves and cycles, and not being fully present and aware in each moment. Your personal vibration drops when you encounter fear and try to deal with it through an unhealthy feeling habit—through fight-or-flight methods. It's easy to get stuck when your personal vibration drops because low frequencies cause more negative experiences. If you try to stop a wave or force it to move as your willpower dictates, you'll cause repercussions and distortions in your life's flow. If you try to leave your experience or focus on emptiness or negative realities, the lack of presence will cause distortions and snags.

The right use of your willpower is not to force, control, or resist, but to: (1) choose a higher vibration, (2) choose to attune to the wave motion you're in and "go with the flow," and (3) choose to "be with" whatever is happening in the moment—to instill more presence into every situation so the wisdom of your soul can be revealed. By dissolving soul-blocking thoughts and the inherited thought overlays that are not appropriate for who you are, you create more clear space for your own diamond light to flow into your body and life. This requires no force—your frequency rises naturally when left to its own devices. It's easier to free yourself from negative vibrations today because the accelerating frequency of your body and the earth makes it hard to remain stuck for long, and clearing fear is much more instantaneous.

Home Frequency Message

As I explain on page xxi in *To the Reader*, I've included these pieces of inspired writing at the end of each chapter as a way for you to shift from your normal, speedier reading mind to a deeper kind of direct experience. Through these messages, you can intentionally change your personal vibration.

The following message is meant to transport you into a way of knowing the world that's close to the way you'll experience life in the Intuition Age. To move into the *home frequency message*, just downshift to a slower, less hurried pace. Take a slow breath in, then out, and be as calm and still as possible. Let your mind be soft and receptive. Open your intuition and prepare to *feel into* the language. See if you can experience the deeper realities and feeling states that come alive *as you read*.

Your experience may take on greater dimension in direct proportion to the amount of attention you invest in the phrases. Focus on the words a few at a time, pause at the punctuation marks, and "be with" the intelligence delivering the message—live and right now—to you. You might speak the words aloud, or close your eyes and have someone else read them to you and see what effect they have on you.

BECOME TRANSPARENT AND POROUS

Imagine seeing yourself as light and energy. Part of you, where you know and practice the many forms of love, glistens with the clear transparency of the diamond. Other parts, where you cling to fear and the past, where you don't experience your true self, appear to be cloudy, foggy, dense, and opaque. In these contracted places, the shadowlike thoughtforms live, the freeflowing energy of the universe's unified field becomes trapped in sinkholes and whirlpools, and this colors your experience of yourself and all other selves since you must look out through this splotchy filter. Seeing through the darkened lens, you experience pain, limitation, lack, and negative emotion and can easily think the shadows are in the others.

The good news is that you are in a process of becoming totally transparent. To be transparent means to hold nothing to yourself, to release all ego, to be soft and adaptable, porous and permeable. It means living without the need for fixed identity or a set history, without limitation, beliefs, fear, or reactionary behavior. When you let go and experience trust or faith, you become more transparent. Whenever you stop being defensive, righteous, aggressive, opinionated, or competitive, your diamond light has a better chance of shining out from you. Every time you let yourself be flexible and fluid, and fully in the moment without expectations or projections into future or past, you increase your ability to receive everything.

There is tremendous energy and consciousness—beyond anything you can contemplate now—floating freely in the Field. If you hold small ideas or contract with doubt and negativity in the face of expanded self-awareness and self-expression, the energy cannot move through you. When it doesn't move though you, you can't know what it knows. It can't help teach you. It can't help you create. When energy in a field intensifies—the earth's field is now experiencing this—you, as part of the field, intensify as well. If you hold on, or hold back, or hold out and resist, your opacity blocks the waves, and you soon begin to shake—as a dog shakes water from its body after swimming—so you can move the stuck, muddy energy, sweep clean your energy pathways, and shift to a comfortable, harmonious attunement to the field you live within.

Where you still hold negativity and contraction—or opacity—you experience suffering and upsets. The more transparent you allow yourself to be, the less disruption you experience. When intense energy flows through a transparent person, it produces a feeling of heightened divinity, enthusiasm, and light. Place your attention on the image of diamond light and feel that glossy clarity in and around you; imagine it extends to infinity. Diamond light, with its innate loving wisdom, saturates every particle, wave, and being; emptiness is an illusion; Tune to transparency's crystalline bell-tone ringing through all time and space, become its intense purity. Put it everywhere inside yourself. Let it absorb and melt away the shadows in its awesome presence. Total clarity comes with total release and total acceptance of all that is.

5

Feeling Your Home Frequency

My arm for a pillow,
I really like myself
under the hazy moon.

Yosa Buson

Lisa, who is recovering from cancer treatment, told me she was reading in bed and realized her body was vibrating, not with the stressful anxiety or electrical buzz that has often beset her during her healing process, but with bliss. Her body was in bliss! No reason; it just was. She directed her awareness down into the unique feeling state, merging with the smooth, happy energy and simply enjoyed what her body was doing—all on its own. "What a powerful healing force this is," she thought. "How could cancer, or a sinus infection, or even a bruise, exist in a reality this awe inspiring? If my body knows how to do this, how did I ever get sick in the first place?" She discovered that night that she loved her own energy, that there was something about the "real Lisa" that was so beautiful and nurturing— she just wanted to bask in herself, to feel her own essence forever. That night, she said, she stayed awake for hours, indulging in feeling her body's communion within itself. She said the state was so memorable, she can easily recreate it whenever she thinks of it.

You have an amazing vibration like this inside you, too—a resonance that conveys your soul's love, truth, abundance, and joy. It bubbles up from your tiny "quantum entities," waving out through your cells and tissues to fill the space around you. It's always there, and it remains dependably consistent. This is your *home frequency*, the vibration of your soul as it expresses through your body. I call it home frequency because it conveys an experience of Home that's as close to heaven on earth as you can get. Your home frequency acts as a compass—when you attune your fluctuating daily *personal vibration* to this most essential, naturally high-frequency energy, your life stabilizes and unfolds with luck, meaning, and enjoyment.

When you sit in your home frequency, as Lisa did, you feel so fantastic and love yourself and life so much that you wonder how you ever lived any other way. The ideas and answers that come out of your home frequency are always just right and further your soul's expression. It's somewhat shocking, really, that it could be possible to lose your home frequency amidst the hubbub of life, but we routinely do. Sometimes, as in Lisa's case, your home frequency asserts itself spontaneously, but more often, you need to seek it, invite it in, and merge with it. Learning to find, relax into, and share your home frequency is the key to transforming your life and entering the Intuition Age.

> Begin to see yourself as a soul with a body
> rather than a body with a soul.
>
> Wayne Dyer

You've Reached a Magical Turning Point

You've been clearing your unhealthy feeling habits and learning to raise the frequency of your personal vibration. As you reach the crescendo of this clearing phase of the transformation process, life can become intense and chaotic and sometimes look hopeless. The old isn't working; you may feel self-sacrificing, unimaginative, and unable to move forward. You've shifted the emphasis just enough from fear to love that your old reality has destabilized and the new reality of your soul is starting to break through. At this point, your life may malfunction and you may have to let go of goals, possessions, people, or parts of your lifestyle. You may lose whole aspects of your identity, your motivation and direction, and your comfortable habits. It's important not to backtrack into more fight-or-flight reactions. What's really happening is that your soul is saying, "You are not this old, limited self anymore. It's time to discover who you really are and what you can do." This is the point where the phoenix lights itself on fire and mysteriously turns to gold. It's where you come face to face with the choice of who you really want to be.

> Fear not the strangeness you feel. The future must enter you long before it happens. Just wait for the birth, for the hour of new clarity.
>
> Rainer Maria Rilke

How do you find your new identity? Should you be like your favorite hero or heroine? Don't worry—you won't need to copy anyone because your new identity will be uniquely yours, and it will fit you perfectly. But you won't find it outside yourself. The answer is encoded in your home frequency, and if you live in your home frequency, your new identity will uncurl like a new leaf. While your outer world has been cracking and crumbling, internally you've been rebuilding and renewing yourself. Your invisible inner infrastructure is almost complete. The magical turning point in the transformation process happens when you stop paying attention to your old world, with all its hectic busyness and clutter, and shift your full attention to how your soul might recreate everything. The challenge here is that from your point of view *within* the hectic busyness, it looks like if you stop or let go, you'll lose everything, fall into a void, and possibly fail and die. This, of course, is the ego's crazy, desperate view, not the soul's. When it seems like you're facing emptiness, you're really about to find yourself again in a new and better way.

It's Time to Relax and Let Go

Life compassionately gives you a connecting link, a phase between the old and the new. Like a ship going through the Panama Canal, you, too, will move through the "locks" of consciousness, gradually changing from a lower level to a higher one. To experience this period, all you have to do is relax. You don't have to know everything about what your future will be or how your transformation is going to work. You can exhale, be less concerned with externals, and stop pushing. You don't owe other people a description of your likes and dislikes, successes and failures, or plans for the future. You can be like your dog or cat: perfectly real, perfectly happy, and perfectly undefined. You're a mysterious force peering out from two beautiful, liquid eyes and radiating out from a happily vibrating body. You can be yourself without maintaining an ego. You're not going to go Poof! and disappear if you let go.

Certainly, you'll face moments when, as endings loom, your ego will precipitate various kinds of clever ploys to keep itself in control. The idea of pausing and "letting the fields lie fallow" for a short time begets cries of: "But I *can't* stop! If I don't do X, I'll be alone," or "I have bills to pay and people depend on me," or "I'll go off the deep end and drown." Your ego will convince you to go back to an old job you've outgrown just to be secure or to become the flailing, out-of-control victim as your new identity. Your ego will always paint letting go as a black-and-white situation in which both options are about suffering: "Either I sacrifice myself by doing something that's no longer right for me or sacrifice everything I have by falling into the void."

> **Letting go is simply about a return to Being.**
> **When done right, it's about centering.**

Letting go is not about sacrifice, nor does it breed lazy inactivity; it's simply a return to Being. It is a shift from an assertive focus on action and results in a softer, intuitive state where you will "be with," notice, and appreciate what's in the moment with you. It's about moving from noise into silence. When done right, letting go is about centering, and it always leads to your home frequency.

When You Stop and Let Go, It Doesn't Mean Loss

To move into the resting and ripening phase between old and new, you don't have to stop everything altogether, and it doesn't have to take a long time. You do need to experience the purity of the pause, though—which could occur in moments, or a day, or a week. Your soul just has to have a chance to emerge into a clean space, to begin saturating your life so you can actually feel it. It doesn't work to stop and impatiently say, "How long is this going to take?"

Margaret recently developed an illness that must be diligently attended to; her parents just died, leaving her in charge of their estate; her marriage has become uncommunicative; she has given up her art because she has nowhere to paint; and her forays into healing others are not drawing many clients. She feels stuck and depressed. She wants a change but doesn't know where to start. Her home is filled with fabulous art, fetishes, and one-of-a-kind collectibles. There is so much, though, that it's difficult to keep clean, and it's gathering dust. Her yard is unkempt, there's an infestation of rats, and the neighbors knocked the fence over into her property and aren't doing anything to fix it. Life is tangling because her ego is trying to avoid letting go and facing a subconscious fear. She's forgotten her home frequency in the complexity of her life. When you feel crowded and cluttered, it's often because you aren't fully "in" yourself, and other people then invade your space, trying to "find you."

I suggested Margaret first take pressure off herself by pausing her internal dialogue that was repeating variations of: I can't, I have to, and I should. In the quiet she could return to her generous, humorous, creative self, make some space, see what might arise from that more pleasurable place, and only do those things for a while. In effect, she could imagine herself as a dented memory-foam mattress slowly returning to its true shape. I also suggested she empty a room of accessories and be with the spaciousness and blank walls, see what feelings and subconscious issues surface, then see which objects want to go back in the room. "But I don't have anywhere to put things because the garage is full of boxes." Our conversation ended with her not quite understanding what I'd been saying, since it didn't relate directly to her concrete problems.

She called several days later to say that she was anxious and couldn't sleep. It became evident that her ego had turned my previous suggestions into shoulds, and she felt she was being made wrong. She had "heard" me negating the idea that she actually wanted to be a *healer*, not an artist, which she had taken to heart after other counselors told her she was "supposed to" be a healer. She held specific ideas of what being a healer looked like, based on healers she'd seen. Here was an issue whose significance had been largely hidden by the other issues she'd described. Under the issue of being a healer was the real issue: Margaret wasn't granting herself the right to do what *she* wanted. She'd been living according to other people's pronouncements and examples, caretaking and being selfless instead of celebrating her originality in a "selfish" way. She was feeling deprived and self-sacrificing and was hovering on the edge of a primal fear of annihilation if she stopped putting other people first.

I suggested again that she give up the images and ideas of what her life *should* look like, go back to "just being," and act spontaneously from there, entitling herself to follow her fresh curiosities and desires one at a time. Then her *own* form of being a healer would arise, if it wanted to, and it might be grounded in art, teaching, and music—the things she'd always been interested in. But first, she needed to stop her old pattern and return to her home frequency.

> If I create from the heart, nearly everything works;
> if from the head, almost nothing.
>
> Marc Chagall

It's hard to grasp that a breakthrough can be about Being when you're in the midst of the drive for action and results. Solutions look like they must be about more doing and having: If I *had* different neighbors. If I *made* more money. If I could *get* enough healing clients. The ego wants a full-blown strategic plan in ten clearly defined steps to be accomplished in a week. Yet without putting the ego on "pause," the soul's magic can't happen. Your home frequency will surface as soon as you stop paying attention to what's not vibrating in harmony with your most childlike, joyful, curious self. You'll start to feel it as soon as you turn your thoughts toward soul qualities. It's waiting for you when you stop. It's in the silence, and it meets you halfway when you walk toward it.

When and How to Drop into Your Home Frequency

Finding your home frequency is really the big turning point in the transformation process. It's one of the best-kept secrets in life that when you think you're falling into the void, you're really returning to yourself—that what you think is going to be empty is actually full, that when you stop the old, the new immediately begins. It is when panic and complexity peak, when you're using willpower to control yourself and life, that several things need to happen:

1. Hit the pause button and suspend your internal dialogue. Shift from the I-have-a-problem state of mind to feeling even the tiniest bit of pleasure in yourself.

2. Enter your body more deeply, calm yourself, listen for the silence, spread your energy out, and take up more space. You're home.

3. Focus on qualities of soul, like cheerfulness, sincerity, innocence, or playful creativity. Look for and sense your core vibration—your home frequency—that's existed since you were a radiant baby. Think of the *you* that you love, the way you feel when you're being loving, happy, and generous. Sink into the pleasure of your own being. Let your home frequency saturate every part of your body, emotions, and mind.

4. Once you've filled yourself with your home frequency, imagine it as a tone, and in your imagination, "strike the tone of your own tuning fork" and let ripples of your home frequency radiate through you and out through the field around you. Give freely to the world.

5. Imagine that your home frequency is reprogramming and retraining your cells while you're enjoying the experience of it. When something fresh and authentic arises from your calm, open space, it will match your home frequency. Whether it's an emotion that relates to your deeper issues, a curiosity, idea, opportunity, or person, follow the urge and engage with it fully.

In Margaret's case, centering into her home frequency might easily bring solutions to her mundane concerns. But this is not just another ordinary period of stuckness; this is a chance to let go and be reborn. She is smack dab in the middle of the transformation from an old reality and identity to a limitless new life. This is true for you, too, because *this is the transformation time on earth* when it's possible to clear away all your limiting habits. Once you let these crisis times be opportunities for permanent betterment, all that's missing is the imagination of what might be possible and the choice to "go for it."

Sometimes You Have to Make Space Again, and Again

Sometimes—perhaps because of our stubborn identification with sacrifice and suffering—the previous sequence of actions must be repeated several times before your new soul-based reality takes hold. I once had an astrology reading from a brilliant astrologer who made a number of definite, author-itative statements about my future, among them: "Your mother will die this fall," and "You will never marry, and if you do it will be unfortunate." I went home in shock and was so upset I ended up crying. I decided to meditate to calm down, and in the meditation, a vision came to me. I saw the upsetting future he had outlined spreading before me into the distance, emerging from me as a cross-hatched grid of gently waving neon green lines in black space—the way computer graphics models are often depicted—with the negative events glowing at various nodules. I realized that just by contem-plating it, I was investing energy in his vision and was adding to its life force, thus helping that reality become fact. What a waste of time!

At that moment the whole grid of lines, my potential future, rotated half a degree—and totally disappeared! Now there was just a velvety black space extending to infinity, supporting me. It was perfectly silent and calm. I rotated back half a degree and the lines, with their negative potential, reappeared. I rotated back out again and—space. I sat with the experience of endless space, and it was much more pleasurable than the dreadful future that had been placed on me like a wet blanket. It did feel a little odd not to have a future, not to have a "plan," but I realized that odd feeling was really *freedom*! I saw that I'd never really experienced freedom before, though I'd always fought for it, and that if I tried to make my future with my limited mind, it would be a mediocre thing. I knew my life would arise out of the velvety blackness without limitation if I'd just stay in the pleas-ure, with its positive expectation and trust in goodness.

So, for months afterward, whenever I realized I was in the "green grid," thinking negative thoughts, I rotated that half degree in my mind and let my plans and overlays—even the positive ones—melt back into space. I practiced being spacious instead of cluttered. My mother did not die that fall; in fact, she miraculously healed from a cancer that appeared suddenly, and she is still, these many years later, one of the most positive and vital people I know. I rapidly cleared through past relationship wounds, my creativity moved in new ways, and I visited foreign countries I'd never thought I'd see. Most importantly, I was at peace.

To know what you prefer, instead of humbly saying "Amen" to what the world tells you you ought to prefer, is to have kept your soul alive.

Robert Louis Stevenson

You Ripen Naturally Like Fruit on the Tree

Basking in your home frequency, you find an uncluttered experience of "being with" life as it is, and in that is the simple truth, which relieves you. Now you become aware of being replenished, that a deeper part of you io exploring all the *superpositions*, or multiple reality options, you have in your waveform state, setting up new experiences for you—but your mind doesn't have to know about it. The forms of your life may or may not dissolve as you take your attention off them, depending on whether your soul needs them or not. They may dissolve and reappear later in an updated version. The point is that you need *yourself* now; the forms will take care of themselves. Sometimes you only need a momentary dip into the pool of soul; sometimes you need a few days of downtime, or even a few months or years of casting around somewhat aimlessly. Sometimes you just need to stop generating negative static and be more deeply connected to what you have. Life never stops completely.

I thought the fruit on my new dwarf lemon tree would always be hard and green, but like magic, all in the right season, the lemons have become plump and juicy and the flesh is getting soft. I can't make them ready to pick; they are taking their sweet time, fulfilling their potential as only they know how. With your home frequency, you don't need willpower to move forward. If you go with the flow of your wave, release what your ego tells you are your only options, and maintain full presence, this time of ripening will first feel like relief. Then you'll feel full of yourself, like a child. You'll lose track of time, while there will also be the pleasure of something readying itself to be born. You may sense that the time of birth is not quite yet, but when it comes, it's going to be wonderful; you trust that your soul is doing a fantastic job providing what's needed. Finally, when you least expect it, the fruit releases from the tree, and there's a surprise arising of a great idea, a plan, an opportunity, an instinctive act, a helpful event or person.

Your Potential Is Limitless

When contemplating the idea of becoming new, it's helpful to know that as you grow and release the ego's need for a special identity, you are freer to draw from the knowledge and experience base of all the lives ever lived on earth, all the experiences that souls have in other dimensions, and every "superposition" possible in the quantum reality. Your identity, and thus your creativity, is truly limitless: what comes back to you after you let go can be quite miraculous and something you may not even be able to dream of right now. Keep your childlike faith and imagination active, because this is how you'll recognize what wants to become real.

Discover What It's Like to Be in Your Body

To find your home frequency, as I already mentioned, you must bring your attention fully into your body, merge into it, and feel what's going on. The reason is that your soul—and thus your home frequency—is saturated throughout your body, and unless your mind is in your body, too, you won't consciously recognize what "soul" feels like, and it won't be real to you. There's a problem, however, and that is due to your body's tendency to resonate, like a tuning fork, and to change its vibration to match the frequencies in your environment. Though your soul's high home frequency is always present, it can be temporarily camouflaged with a variety of lower vibrations, such as worry, panic, envy, anger, or mental exhaustion.

When the vibration of your body gets chaotic from too many distracting surface vibrations, your mind has a tendency to jump out and project into other time periods and places, or just "space out" into blankness. When there's too much stimulation, or the vibrations are low and negative—if you're not doing what you love, are caught in unhealthy feeling habits, are sick or in pain, or have trashed yourself with drugs or bad food—your body is just not a comfortable place to be. You can probably recognize when someone is "out of their body" with that nobody's-home, vacant stare, when they seem to be automatically reacting or going through the motions instead of being fully involved. When people are out of their body you can't feel them, and you may become anxious: "What are they going to do next? This car has no driver!" It's not unusual to leave *your* body, too. Try making eye contact and smiling at the checkout clerk next time you're at the market and watch them come back into their body and connect with you.

So what is it like to really "be in your body"? Most of us think we "have" a body, that it's down below us somewhere doing something—heart beating, blood moving, food digesting. But to actually be in your body to such a degree that you become the body and know the world from its point of view—that's another thing. When you're in your body as your body, you're like an animal—you're at "eye level" with life. You have *direct experience* of life and a live connection to the world. You respond immediately to situations without pausing to analyze and compare, and you're honest and fully engaged with each action. You're so in the moment that you can instantly expand and contract, go left or right, and adjust your energy level perfectly to fit the circumstances at hand. Some people call it "being in your groove" or in the "zone." When you're in your body fully, your eyes shine, your meaning is instantly conveyed, you're totally convincing, and if you're in your home frequency, too, your body feels safe, healing, and trustworthy to others.

> You are never more essentially, more deeply, yourself
> than when you are still.
>
> Eckhart Tolle

Calm Your Body to Feel Your Home Frequency

When you bring attention into your body and take up residence in the center of the here-and-now, you may notice a surface vibration that's rough and buzzy if you've been experiencing anxiety or pain or if your mind or emotions have been racing. As you merge further into your body, you may encounter more surface vibrations with textures of energy that feel like sandpaper, gummy waxiness, ashen dryness, damp cold clay, or prickly high-voltage electricity. Press on, looking for even subtler vibrations. Even if you think you can't feel anything, drop in further and hold your attention still. You'll discover your nerves are tingling and your cells are vibrating. Your body is going about its business of staying healthy. If you listen closely, you might even hear your body's happy, healthy tone.

Deliberately calming your body helps you drop through distorted surface vibrations faster. Calming yourself is often just a matter of shifting out of left-brain, linear, analytical, verbal perception. By moving to the right brain and artistic, beauty-oriented, intuitive perception, your energy softens and opens. It's more relaxing. Calming your body and raising your personal vibration can also occur when you perform continuous, smooth movement, like walking, running, tai chi, swimming, or bicycling. In addition, it's calming to perform repetitive actions, such as rocking, tapping on your body, breathing consciously, saying mantras, paddling a kayak, or drumming.

Try This!
Enter Your Body in a Deeper Way

Right now you have an opportunity to Be. Bring your attention inside your skin, and forget that there is such a thing as a past and a future. This, right here, is fascinating.

1. Say to yourself: "In this moment and in this body I am 100 percent present." Feel what it means and come into alignment with the statement.
2. Turn down the sound of your thoughts as if you're adjusting the volume on a radio. Soon you'll hear your body's teeming, which might sound like ringing, white noise, or a steady hum. Imagine something deeper behind or through the body hum. Enter the silence at the core of yourself, which is always there in spite of any physical noises.
3. Imagine you're in the center of your head looking out from behind your eyes. In that centermost point is a tiny gleaming diamond, emit-

ting transparent light that radiates through your brain, clearing your mind to a neutral state of observation.

4. Imagine you're inside that diamond looking out, and it can move through your body like a tiny flying saucer. Let it fly down to your throat and hover. Look out at the world from this vantage point—your body's head is up above you now. Then fly down to the center of the chest near your heart and hover. Look out at the world. Some of the body is above you, some below. You're centered in the middle.

5. Now fly down to the base of the spine and hover. Look out at the world from this vantage point. You're much closer to the earth's energy, the mind-in-the-brain is far above, and the body understands other bodies directly without language.

6. Experiment with flying to a variety of places in your body and feeling the vibrations in the arch of a foot, a kneecap, the tip of your pointer finger, the base of your tongue, the center of a vertebra, your heart, your diaphragm. As you take these various vantage points, you may notice that you know the world in a particular way, that there is a certain kind of awareness inherent in each place. Some places are incredibly quiet and wise.

7. Come back to the center of your head, open your eyes, and walk around, paying attention to your environment by noticing only color, shape, texture, temperature, smell, and noise. Don't label anything; just remain in the direct experience, moving smoothly from one impression to the next, as an animal might.

8. Later, try doing an activity that emphasizes one or two of your senses: dance around the living room to music, make a blender drink with fresh ingredients and drink it slowly. Notice your body's pleasure, and note exactly how it feels.

Your Heart Is the Key to Your Home Frequency

Perhaps the most calming way to find your home frequency is to enter and activate your heart awareness and apply loving attention to your body and the situation at hand. A quick way to do this is to *allow* life to be the way it is, choose to find the soul's sanity in your immediate experience, and remember how you like to feel when you're at your most generous, kind, and unselfish. Be that way toward the moment and all it includes.

Everything, in some way, is in perfect order; let that purposefulness be revealed to you. Be welcoming.

Home Frequency and Your Heart's Electromagnetic Field

In an article in the Winter 2005 Institute of Noetic Sciences magazine, *SHIFT*, Rollin McCraty, Raymond Trevor, and Dan Tomasino wrote about our body's heart-field. They said that the "heart generates the body's most powerful and most extensive rhythmic electromagnetic field." It is sixty times greater in amplitude than that of the brain, permeates every cell, and can be detected several feet away from the body. "Negative emotions, such as anger and frustration, are associated with an erratic, disordered, incoherent pattern in the heart's rhythms." Positive emotions correlate with a smooth, coherent pattern, and "sustained positive emotions appear to give rise to a distinct mode of functioning, which we call *psychophysiological coherence*." This mode is characterized by "increased efficiency and harmony in the activity and interactions of the body's systems," as well as a "reduction in internal mental dialogue, reduced perceptions of stress, increased emotional balance, and enhanced mental clarity, intuitive discernment, and cognitive performance." It makes sense that when your heart's vibrations are "coherent," your body, mind, emotions, and soul are functioning in harmony, or perhaps at octaves of the same frequency—your *home frequency*.

I've seen an amazing thing happen in people when their heart opens: their eyes widen and glow, their face smiles without restriction (as my little sister used to say, "Help! I can't stop smiling!"), understanding floods their mind, and their *third eye*—the energetic center or chakra in their forehead—opens, bringing sudden insight and visions. If you can apply compassion and tenderness to the fear that's making your body shake, or to the contractions causing your physical pain, and if you can talk soothingly to your body as though it's a child who's fallen off his first big bicycle, you'll enter the heart-field. It's this particular vibration, which Buddhists often call "centered dignity," that will help you discern your home frequency.

Try This!
Feel into Your Body to Sense Your Personal Vibration

This week, at least once a day, center yourself by bringing your attention inside your skin, focusing on the electromagnetic center point in your brain, and being receptive to whatever you notice.

- Feel into your tissues, organs, bones, and cells. What does the vibration feel like? Is there a surface vibration that feels overactive or underactive? Describe it in your journal with sense-oriented adjectives: What does it sound like? Look like? Kinesthetically feel like? Taste like? Smell like? Are there any emotions you'd naturally connect with the vibration?

- Feel through or beneath any surface vibrations by going to the innermost places in your body and into the core frequency of your heart. What does this steady vibration feel like? Describe it in your journal with sense-oriented adjectives: What does it sound like? Look like? Kinesthetically feel like? Taste like? Smell like? Are there any emotions you'd naturally connect with the vibration?

- Over time, notice your habitual ways of covering over your home frequency with surface vibrations and see if you can catch yourself being "off-key," then practice surrendering back into your core. Don't look for a change of form, just enjoy Being.

Your Home Frequency Is Not So Much "High" as It Is "Real"

It's true that your home frequency is very high, but it would be a mistake to think that you get to it by jacking up your vibration or "trying" to be high-toned. If you use will to try to have or be something, it means that underneath, you feel you don't already have it or aren't naturally that way, that there's a gap to cross or an obstacle blocking the experience. The more you try, the more shrill and brittle your vibration becomes, and it takes you further from your naturally high home frequency. You don't have to generate your home frequency; it always radiates freely. Just relax into it. The more honest you are, the easier it is to stay tuned to your most-real self. *Your personal vibration may vary from minute to minute, but the goal is to keep it attuned to your home frequency so that your personal vibration eventually becomes your home frequency.*

Like Lisa, you may recognize your home frequency through a specific soul quality, like bliss, cheerfulness, generosity, sweetness, joy, sincerity, dignity, or readiness to laugh and be involved. Or it might be a texture of energy that feels like silk, soft butter, diamond, alpine air, or running water. You may feel it one way when you're quiet or meditating and a different way when you're in action. It can help to zero in on your home frequency through a process of contrasting the old, slower frequency states you've assumed were normal with higher states you may recall only vaguely. The goal is to take a snapshot of your home frequency state in tactile detail, so you won't lose track of it. You need a reliable feeling to center back into when you're knocked off balance by agitated or wounded people, scared out of your body by experiences that remind you of past traumas, or overwhelmed by too many choices.

We must be still and still moving / Into another intensity . . .

T.S. Eliot

What's Your Worst Possible Scenario?

Imagine you are in that state of mind where you're preoccupied with troublesome issues and problems. Perhaps other people are bothering and distracting you, or you feel too crowded with concerns to think clearly. Maybe you're worried about failing or a loved one's health. There's a mood we all fall into that promotes anxiety and agitation. See if you can call it up and feel it temporarily.

Try This!

List Your Negative Preoccupations and Worries

1. Tap into your "worry mood" and list things that fall into the following categories:

- What problems are you currently trying to solve?
- What issues are you working on in your personal psychological and spiritual growth process? What old wounds seem to be surfacing?
- What things related to environment and other people have been bothering you lately?
- Who are you concerned about and why?
- What's frightening you or making you anxious?
- What produces chaos in your life right now?
- What do you physically need and desire right now that you're not getting?
- How do you feel deprived emotionally?
- What are you resisting? Where is the conflict in your life?
- What things make you feel contracted and tight when you think about them?
- How do you feel trapped? How do you feel overwhelmed?
- Where in your life do you feel something is unfair?

2. By fleshing out and recalling specific instances where your personal vibration contracts, where worries or densities pile up on one another, you can easily feel your low-vibration state of being—what I like to call our "old reality." Next, pick a few of the items you listed above and write about the specific sensations you experience in your body as you are tuned to the negative vibrations. Feel the reality.

3. From the state of tension and anxiety you created by listing and feeling your limiting conditions, project your life into the future and paint a picture of your worst possible scenario. What might happen if everything goes wrong and you have bad luck and nobody helps you? Imagine it in detail and take a mental snapshot so you can remember it. Then exhale and relax.

By stretching your slingshot back to the furthest, tightest position, when you release it, you'll go the farthest distance into your positive reality. The point here is to consciously recognize how awful the negative reality feels and to pinpoint specific sensations in your body that go along with this vibratory state. You may notice a tightening of your throat or chest, a clammy coldness, or a tendency to hyperventilate. Maybe you feel heavy and dark inside or want to explode like a nuclear blast and flatten everything around you. These tendencies relate to your methods of unconscious stress management.

What's Your Best Possible Scenario?

Now let's shift gears and contemplate the things you appreciate or helpful things you'd like to do for others. Perhaps you're dreaming about a fun vacation, a play you're going to see, or a creative project you're about to undertake. Maybe you're feeling happy for a friend who's had some good luck. There's a mood we all fall into, quite opposite from the previous one, that promotes enthusiasm and flow. Call it up and feel it.

Try This!

List Your Positive Experiences

1. Tap into your "happy mood" and list responses to the following categories. As you connect with memories and experiences, feel your body shift out of any residue of contraction left from the previous exercise and into a resonance that easily calls forth your soul qualities. Recall one or more specific experiences when:

   • Your gut feeling led you to a great decision.
   • A fabulous opportunity fell in your lap or someone was amazingly generous to you.
   • You met special people who felt like soul mates, soul friends, or soul family.
   • You were generous without concern for repayment or acknowledgement.
   • You heard important truths and information that excited you to new growth.
   • You were deeply quiet, peaceful, satisfied, and grateful, and felt that "all is well."
   • You received a vision and successfully made it happen to your own standards.
   • Your imagination flowed and wonderful, original ideas came to you magically.

- You experienced a period marked by synchronicity, luck, flow, and cooperation from others who were talented.
- You received just what you needed without having to ask for it.
- You were in such a good mood that no one could ruin it for you.
- You were with a special animal or in a special place in nature, feeling connected to life and perhaps to a sense of the Divine.

2. By fleshing out and recalling specific instances in which your personal vibration expands, when successes and good luck build to greater possibilities, you can easily identify and feel your home frequency and the "new reality." Next, pick a few of the items you listed above and write about the specific sensations you experience in your body as you are tuned to the positive vibrations. Feel the reality.

3. From the state of enthusiasm and centeredness you created by listing and feeling the positive experiences you've known, project your life into the future and paint a picture of your best possible scenario from today's vantage point—what might happen if everything goes well, you have great luck, everyone helps you, and all the positive qualities naturally increase and your capabilities expand? Imagine it in detail and take a mental snapshot so you can remember it. Then exhale and relax.

Again, the point is to consciously recognize how fabulous and uplifting the positive reality feels, and to pinpoint specific sensations in your body that go along with this high vibratory state. You may notice a spreading warmth, a feeling of relief, energy bubbling up, an expansion of your energy field, or a desire to beam or shine out. Maybe you feel generous, loving, and tender. Perhaps you feel excited, like a racehorse in the starting gate.

Try This!
Rock Back and Forth Between the States

You now have two well-defined states of being to choose between. It's like a prize fight with two contenders. In this corner, the illusionary vibratory state from hell, which tricks you out of feeling your soul! In the other, the true home frequency from heaven that reveals who you really are! You've been rocking back and forth between them randomly, being first one then the other, but now it's time to take charge of your rocking motion so you don't stay too long in the world of fear and suffer needlessly.

1. From your current vantage point of just having been in your home frequency, try rocking to the opposite side and merge back into the contracted, self-sacrificing, worst possible scenario state. Let yourself remember your list of problems and upsets and sink back into that way of being. Notice how it feels.

2. Now relax and rock back to your home frequency. Review your list of positive experiences and let yourself shift and attune to these higher vibrations. Notice how you feel and the changes in your body.

3. Now rock back to the contracted vibration and attune to it, merge with it, and become it. Notice how your body feels. Does it like this state?

4. Now rock back to your home frequency. Feel the openheartedness, the relaxation, the naturalness, the fluidity, the permission. Does your body like this state?

After rocking between the extremes of worst and best possible scenarios, you'll notice that going back into the contracted, denser state takes some work. After a few times, your body is probably saying, "But do I *have* to?" It's not easy to shut down and maintain that level of tightness and falsity, yet of course we do it all the time. In contrast, you can feel how effortlessly things work when you're in your best possible scenario state, attuned to your home frequency.

It's Up to You to Choose How You Want to Feel

Are you waiting to feel good? Every day, in small situations or large, you have the choice to be either contracted and anxious or at-home-in-the-center. It's up to you how you want to feel and who you want to be. No one else can create the conditions for you to feel good if you haven't decided to feel good. There comes a point in everyone's life when we must simply decide to be fine, to feel healthy, and not to wait for situations to clear up or problems to be solved. Life is short; we might as well enjoy it. And enjoying each moment, one at a time, is not such a difficult task.

This is where we must be brutally honest and ask ourselves the withering, humiliating, yet liberating questions. Am I being miserable to punish others? Am I stalling my growth because I stubbornly want others to make me feel safe or do things for me? Am I hesitating to have a good life because I don't want to admit I've made a mistake? Am I maintaining my pain because I'm too lazy to think of anything better? Am I proud of myself? The biggest question of all is: am I willing to let go of the old, transform, and know how truly good it can be when I'm not in control?

As you contemplate the answers to these questions, notice what your excuses are. How do you explain your lack of courage to yourself and others? We all have justifications for why we don't live fully. Yet each time you recite yours, it actually generates shame at a deep level, making it harder to let go.

It's also helpful to realize that this very body that we have, that's sitting right here right now . . . with its aches and it pleasures . . . is exactly what we need to be fully human, fully awake, fully alive.

Pema Chödrön

Look into your life and see to what or whom you've given the power to displace you from the sanctuary of your home frequency. What have you allowed to be more important than your inner peace? Is it the negativity of politics? Or an illness, a crazy boss, inconsiderate drivers, barking dogs next door, an alcoholic sibling, or not getting a thank-you note? If you want to proceed into your new high-frequency life, you'll need to take your power back from all these things. That means when they arise, you let them be and choose your home frequency instead of your "upset frequency." You focus strongly on the nuances of how your home frequency is positively affecting your body, emotions, and thoughts.

As you become clear about living in your home frequency and returning to it promptly the minute you notice you've left it, you may make some statements to yourself to help you remember that your home frequency reality is the *only* reality. They might include: I will stop passing along my pain. I will not say or do things that cause other people physical, emotional, or mental pain. I will not think about myself or others in ways that reduce our magnificence or potential. I will look for the good reasons in my experience. I will be compassionately honest. I will trust and act from my soul in each moment. When you intentionally choose your home frequency, you're doing nothing less than catalyzing your enlightenment and that of everyone else.

Your Diamond Light Body Is at the Wheel

To find yourself and maintain your frequency when you're in an unfamiliar context, it's important to remember who you are. You are a limitless being made of consciousness and energy, connected to the all-that-is through a unified field that coordinates and regulates the flow of creation *perfectly*. Your basic substance is light. When I first learned to do intuitive life readings for other people, I was taught to silently say to my client: "The light in me knows and loves the light in you." By feeling the truth of this, an important attunement was immediately created. Over the years, I've come to see people's soul energy as transparent, diamond light. It holds nothing, has no blocks, and is brilliantly pure and clear. You can use the following powerful meditation to help you find your home frequency when you lose it. With practice, you'll learn to do it in seconds—before a meeting, when you're giving a talk, or even when you're driving.

Try This!
Activate Your Diamond Light Body

1. Quiet yourself, center your attention in your body, be in the present moment 100 percent, and create calmness and receptivity. Recall your list of positive experiences.
2. Imagine that behind your back, your diamond light body appears. This body looks just like you but is made of pure transparent light and has no wounds or blockages. Your light body radiates wisdom, love, harmony, and the knowledge of abundance. Your light body steps forward and puts its hands on your shoulders.
3. In your imagination, feel the higher vibration of your light body; welcome it and attune to the frequency. As you do, your light body steps inside you, merging with you seamlessly and easily.

4. Your light body matches up with your physical body perfectly, each light body part finding its physical body part: the light heart merges with the physical heart, the light cells match the physical cells, the light brain joins the physical brain. Take some time to scan through the various parts of your body as this process occurs.

5. Allow yourself to let go into the light body as it takes over, saying, "You know how to run this brain, this heart, these lungs, how to use these hands, this voice. Please show me how. I trust you to renew me, reorganize me, and teach me." Fall into your own light and feel supported.

6. An odd thing happens; as you let your diamond light body take over to guide you, a saturation point is reached where you "flip" and realize you are the diamond light body. Your identity shifts. As you hear the voice of guidance from the light body, you realize it's your voice. You might say, "I am here now, and I know what's real."

7. Let the diamond light saturate not only every cell, but your emotions, feelings, and thoughts as well. Let it work on your brain and body, dissolving shadows, filling in gaps, upgrading all your systems, erasing worry and doubt, opening new pathways, and reprogramming you with updated frequencies. Remain in the silence.

8. Now, "strike the tuning fork" of your diamond light body's vibration and let waves of your light and your original tone ripple through every tiny space in your body and out through your skin into the space around you. Let it expand as far into the universe as it wants. As your own diamond light expands, it joins with the diamond light it encounters in the presence everywhere. In the center of the light, you can hear or feel your soul's eternal indestructible tone or home frequency.

Living in your home frequency clarifies your perception. As you become calmer and more centered, others around you may be hitting the earlier, more chaotic parts of the transformation process. You may be tempted to merge with those who are suffering to try to help them. You will probably be knocked out of your center and into lower frequencies a hundred times a day. You'll have to be enormously kind with yourself, yet disciplined to "choose once again." As you become accustomed to living in your home frequency, you may find others coming to you to discover something about that "special quality" you have.

Cameron, a management consultant and martial artist, described how the company he works for, which was founded by two visionary men, was recently bought by a large tech company. The new top-level manager is impersonal, hates managers, and withholds important information. The result is that no one knows how to track their performance, set goals, and work out problems. The company has morphed from a fabulous place to work into pure bedlam. Cam, with his knowledge of energy states and flow, has been watching the shift and the patterns happening below the surface. He said he does not manage any of the others, but because he is generally centered and calm and keeps his heart open during problematic situations, people are coming to him for guidance and much of his time is being eaten up mentoring his colleagues.

It's interesting to see that it's the inner energy reality that people respond to; Cam is not the leader of the group, but his energy is the most trustworthy. Yet he, too, is feeling challenged, being surrounded by a field of chaotic energy every day. It's worth it to him to stay with the job a little longer because he feels he's practicing maintaining his home frequency, much as he practices his Aikido, but he also feels his talent is being wasted on the "energy games" that are being played. At some point in the near future, he anticipates finding a new situation where the people are more on his wavelength.

> The privilege of a lifetime is being who you are.
>
> Joseph Campbell

Just to Recap . . .

You've reached a major turning point at which you're ready to release the old fear-based reality and move into the soul-based reality. To do this you need to stop, be present in your body, and find your home frequency—the vibration of your soul expressing through your body. Your willpower and ego cannot control everything. It's time to "just be" and "be with" life as it is. Letting go allows you to shift into a state of fullness in which you sense something authentic ripening inside, yet you can't rush it. All in due time, and when you least expect it, perfect solutions and new ideas spontaneously appear, carried along by the enthusiasm, trust, and positive expectancy of your home frequency. By entering your body more deeply and calming it, and by making space and dissolving clutter repeatedly, you'll be able to drop through distracting surface vibrations more quickly.

Your home frequency is not found by jacking up your vibration to an artificial "high" but by being real and honest in each moment and letting your energy radiate naturally, without interference. You can find your home frequency by opening your heart and practicing lovingkindness. It also helps to contrast the felt sense of your worst possible scenarios with your best possible scenarios, to rock from the tightest, most contracted state into the most open and fluid one. This way you can pinpoint your home frequency state and return to it when you are knocked off balance by lower vibrations. You need to be honest with yourself and determine who you want to be and how you want to feel. No one else can do it for you, and there are various excuses and stalling mechanisms you may need to dissolve.

Home Frequency Message

As I explain on page xxi in *To the Reader*, I've included these pieces of inspired writing at the end of each chapter as a way for you to shift from your normal, speedier reading mind to a deeper kind of direct experience. Through these messages, you can intentionally change your personal vibration.

The following message is meant to transport you into a way of knowing the world that's close to the way you'll experience life in the Intuition Age. To move into the *home frequency message*, just downshift to a slower, less hurried pace. Take a slow breath in, then out, and be as calm and still as possible. Let your mind be soft and receptive. Open your intuition and prepare to *feel into* the language. See if you can experience the deeper realities and feeling states that come alive *as you read*.

Your experience may take on greater dimension in direct proportion to the amount of attention you invest in the phrases. Focus on the words a few at a time, pause at the punctuation marks, and "be with" the intelligence delivering the message—live and right now—to you. You might speak the words aloud, or close your eyes and have someone else read them to you and see what effect they have on you.

MERGE WITH THE HEART-FIELD

All around you: feel the restful, blue-black velvet of deep space. It supports you as the sky supports the stars. Floating in the center, entirely trusting the cradling silence, you are being created moment-to-moment by the power of a wise Presence that never abandons you. Its gift of constant loving attention is the gift of life, the gift of self, and it is the essence of Heart. The surprise is: inside your own heart is the Presence that creates and lives in every heart. And you are a presence that lives inside the great Heart.

Imagine: your relationships dissolve, your work dissolves, your possessions dissolve. You don't need food, water, money, or approval. You have no goals, no needs, no body even. Everything solid and slow, gone into transparency. You have no beliefs, no right-and-wrong, no memory, no growth, no mistakes, nothing to communicate. What's left is a vibration of clear light, a wave of love. You are glowing, floating, beating heart-stuff. This is who you are, and how you know, and what you know.

Look now across the cosmos-field; each bright star is a heart made of light and Presence. Look up close in the earth-field: each bright light is the heart of a being who lives or has lived. One is the world leader lost in power, one the starving mother and one her dying child, one is the ascended master, one is the waitress from lunch, one is the soldier who won't come home, one is the angel guarding him, one is your favorite pet, "passed on." Look closer, in the body-field: each bright dot is a tiny cell-being, a micro-heart receiving and giving light, relaying love, a piece of fairy dust magically shaping life.

When you see the distant star as a heart, it sees you as a distant star with a heart. When you see the immortal heart-life in the dying child, she gratefully celebrates your eternal heart. When you see the loyal cell as a heart, it sees you as its magnification. These lines of recognition are the vibrating love-strings connecting the conscious cosmos; these strands of attention weave us into a glowing resounding heart-field; these filaments of energy are the recognition of our oneness. Picture and feel the enormity of the web of hearts and heart-lines in the heart-field. Imagine the steady, refined, life-giving, love-generating tone reverberating through it. It is the sound of "home."

Imagine: all the hearts are looking at you and know you. Receive the loving attention unconditionally and pass it on as you place your conscious attention on others. Receive it and give it, receive and give. The heart-light soaks into you and you grow and grow; you remember your self, then the wave moves on, overflowing, cascading, and rippling through space. Now more heart-love comes to you. Receive and give, receive and give. The Heart is beating through the heart-field, reminding you, with each wave, of the Presence within us that supports us all.

6

"Feeling Into" Life with Conscious Sensitivity

Innovative leaders working in the invisible realms act on their subjective perceptions of the organizations or situations... Such leaders work with the invisible, subconscious mind to foresee impending chaos... They are able to turn chaos into creativity and to support the unquantifiable shifts in thought that must transpire in order for a transformative change to occur.

Karen Buckley and Joan Stuffy

Y ou, as the newly risen phoenix, now face the prospect of reentering the world and learning to adjust yourself to a new "normal" reality where surprising things may happen. You may routinely experience synchronicity and telepathy, take right action before you consciously realize what needs to be done, know things without going through the usual channels, and rely on subtle energy ebbs and flows to help determine your direction. You may be much more sensitive than you were before, and you may be wondering how you can harness this new, relatively unfamiliar power of awareness.

Sensitivity is the ability to discern physical and emotional feeling and sensation, and as it heightens, you may pick up nonverbal information through your five senses and via tiny variations of expansion, contraction, and movement in your body. *Conscious sensitivity*, which you're about to explore and practice developing, is the ability to immediately perceive subtle stimulation and nonverbal information from either physical or nonphysical sources and to discern the meaning *as it occurs*. As you become more skilled at increasing your ability to know directly via vibration, you'll find that there are many improvements in your life and that this new power of awareness can be used for things you never imagined possible. At this point, though, you may be asking some very basic questions: How do I engage with the world? How do I navigate and know what's meaningful?

"Feeling Into" Is the New Rule of Engagement

Engaging with the world is a new kind of proposition now. It's not so much about ambition, attracting attention, conquering, or making deals face to face as it is about sensing your way into a shared experience with people, objects, machines, processes, and events. Direct experience (conscious sensitivity), direct knowing (intuition), direct communication (telepathy), and direct loving (heightened empathy)—all functions of your home frequency and the present moment—will be your primary methods for navigating, knowing, and acting.

Here's how it begins: You center in your home frequency, then intentionally extend your feeling sense 360 degrees around and beyond your body. As you do, you include things and notice certain things in particular. Then you *feel into* whatever you notice. Feeling into is using your sensitivity to penetrate into something, merge with it, and become it briefly. As you feel into something, it becomes part of you, and familiar, and you begin to know about it immediately. It's a little like role-playing the things you're noticing.

> Beyond my body my veins are invisible.
>
> Antonio Porchia

Let's say you expand and notice your sofa. You feel into it. By merging into and becoming it briefly and acknowledging its unique kind of consciousness, you know firsthand that it wants new foam for its cushions. Then you expand some more and notice the tree outside your window. Feel into it, and by being in its place and living its life momentarily, know it's being crowded by another tree and wants more light. Now you expand and notice the bus on the city street, now the group meeting at city hall, now the valleys or mountains a few hours away, and so on. Whatever you take time to feel into reveals its inner nature. This happens by a kind of osmosis, as you become and "be with" the thing you're paying attention to. By feeling into the world around you, you feel more related to everything.

By knowing and living in your home frequency, you establish it as a reliable baseline for comparison to other kinds of frequencies. After you feel into the sofa or the tree and try on their vibrations and viewpoints, you can easily snap back to center and remember your own frequency again. The next step, after comparing vibrations consciously, is to use the changing state of your body and personal vibration as a barometer of what's happening around you. Your body, remember, is like a tuning fork and easily changes its resonance to match what's in the immediate environment. As you meet a highly animated person, for example, you may also become excited. Your body is registering data by matching the other person's vibratory state. But when you recenter into your home frequency, suddenly you know whether the person is a genius with a great new idea or a bipolar person who's in need of medication. Your body becomes a vehicle for receiving vibrational information, and your home frequency helps you decipher it.

Sometimes you'll receive vibrational information without being aware that you've been observing anything. You suddenly get the urge to call your sister, and it turns out she's having a crisis at work and could use your input. Because your body is always on the alert for potential dangers, you may pick up stray negative vibrations that aren't particularly pertinent to you but which cause you undue stress. For example, you may notice you feel worried for seemingly no reason. By recentering in your home frequency, you realize that tension arose earlier in the morning because during your commute, people on the freeway were unusually angry. This is not an immediate danger, so you can let the insight go and relax. When working with sensitivity, it's important to stay attuned to your home frequency and ask yourself often throughout the day: "What am I noticing? What do I already know about this? Is this information useful and appropriate for me now?" Build a habit of checking in, listening to your body, and downloading important sensory communications, just as you would read your email and throw out the spam. By making vibrational information conscious, you clear your "screen" and minimize needless distractions and negativity. And you increase your conscious sensitivity bit by bit.

Conscious Sensitivity Helps You Know Things Firsthand

When you let go of preconceived ideas, focus your attention and sensitivity on something, feel into it, and be with it—especially with a true desire to know and appreciate it—amazing knowledge is revealed. By feeling into and temporarily merging with the life and viewpoint of a person, animal, plant, place, problem, or situation, you can learn the secrets encoded there through direct experience and direct knowing. Feel into a plant and you know it needs water or fertilizer because you can feel that you need water or fertilizer. Feel into your pet and you know it wants to go for a walk because *you* feel like going for a walk. Feel into a place and you know it has underground water or mineral deposits because you can feel those resources as a part of *your* body. Feel into a problem and you know it wants to naturally evolve in a certain direction because *you* feel inclined to go in a direction without being restricted. Feel into your mate and you know he needs to get something off his chest so his heart can reopen, because *your* heart feels tight and *you* want to talk.

You'll know what to say *yes* or *no* to by how deeply comfortable or uncomfortable you feel. When you feel into a process or situation—by paying attention just a little longer, then a little longer—you'll receive layer upon layer of data about how things are likely to unfold because everything contains its own full story if you penetrate into it far enough. Feeling into life is like doing your homework before you commit to action. It's after this that you can use your strategic mind to investigate and carry out the insights gained. You'll still need to balance your checkbook and analyze data, but logic and organizational thinking will be used more for implementation now.

Until you build your sensitivity expertise, just feel your way through close-up, mundane choices. Even if you have big decisions looming and putting pressure on you, don't jump ahead and overwhelm yourself. The first step toward the big decision is a small decision. Practice sensing whether you want eggs or oatmeal for breakfast. Then feel the exact right moment to leave the house for an appointment. Then feel which phone call to make first. Make these choices consciously by observing how the flow of energy wants to move you and what gives you a feeling that best harmonizes with your home frequency. By the time you reach the big choices, you'll be skilled at gathering the insights you need by reading the vibrations around and within you.

> Even if our efforts of attention seem for years to be producing no result, one day a light that is in exact proportion to them will flood the soul.
>
> Simone Weil

Through *direct experience*, which is feeling and sensing without the mind's commentary, you'll pick up impressions that may seem simplistic but are loaded with meaning. For instance, you might interview for three job opportunities. Which one is best? Your body responds to the first interview by feeling cold, to the second by getting sleepy, and during the third, you smell fresh flowers and are unusually articulate. You intuitively like the third option because it feels welcoming and natural. When you walk to your car afterward, you feel slightly sad to be leaving the building. More research may be required, but you can trust your impressions; your body doesn't lie.

You might be with a client and telepathically hear your own inner voice complaining about not being listened to—and suddenly realize this is what your *client* is thinking. Immediately you say, "Am I hearing you right? Is there something I've perhaps misunderstood?" You might find that heightened empathy— the sense of loving the light in others and feeling connected because of it—helps you know exactly what people need, even before they do. You may know, without having a reason, that it's time for your family to change their diet and eating habits, or that a friend is lonely and needs a fun outing.

Try This!
Feel into an Object, Machine, or Plant

1. Pick an object. Place your attention on it. Keep your awareness open to any impressions from it. Grant it life and let yourself be curious to get to know it, as you might a new best friend. Begin with your sense of sight and examine its looks. Then shift to your sense of touch and— without actually touching the object—expand your light body beyond

your physical body, letting it be a field of malleable energy, and let part of it move like a cloud toward the object and envelope it. Imagine that your eyes are in the cloud. Imagine getting up very close, almost at the molecular level, so your particles of light can move into the object's particles of light. Your point of view is right down there next to the matter of the object.

2. Flow into it with your energy and awareness. Greet the consciousness of the object and ask if it's OK to share its space for a while and get to know it. As you merge with it, sensations, impressions, and insights will be transmitted instantly into your body and mind. Stay relaxed and curious. You may receive impressions about the object's history, life span, potential, symbolic meaning, structural strength and weakness, and needs.

3. Make mental notes. Radiate a vibration of appreciation and love into the object, expressing gratitude for its existence. Then pull out, come back fully into your body, and review what you've noticed and learned.

Try This!

Feel into an Animal or Another Person

1. Pick an animal or person, present or at a distance, and go through the same process as above, making sure to ask permission to share their space and get to know them. You may receive impressions from a living creature concerning health, emotional needs and habits, talents and desires, energy flow or stuckness, thought patterns, and destiny.

2. Before withdrawing your light body, thank the being's body and soul by radiating appreciation and heart energy and conveying the thought that you're happy they're alive. Come back into your body and review what you've noticed and learned.

Feeling into Life Creates Intimacy and Caring

When you feel into life, you learn to care more about everything, and what you care about reciprocates. As you encounter something that your mind might label as "not-me," like the kitchen table, if you don't leap into your concepts and label it a "kitchen table" (which keeps you "up in your head" and separate from it), you'll directly experience it as a kind of energy that's part of you because it exists in your *personal field*. Your personal field is the subtle energy inside and around your physical body containing your emotional, mental, and spiritual awareness. Anything you notice immediately becomes part of your personal field and, therefore, part of you. So if you feel into and merge into the form of the-object-that-shall-remain-nameless, you'll have perceptions that originate from its point of view. You'll feel its life force, harmony, beauty, love, integrity, and health. You'll know what it's like to be wood or steel, for example, or to have four legs or hard or rounded edges. You forge a bond with what was previously foreign, even if it's an object or a machine. I regularly take time to feel into my car, computer, washing machine, and hair dryer, and love them, and I'd swear that my machines look out for me and rarely malfunction because of it. They send me thoughts: "Please clean the lint filter," "I need an oil change," "I'm going to break soon so don't take me on your trip."

When we are really awake to the life of our senses—when we are really
watching with our animal eyes and listening with our animal ears—
we discover that nothing in the world around us is directly experienced
as a passive or inanimate object. Each thing, each entity,
meets our gaze, with its own secrets . . .

David Abram

By learning to consciously merge with people and things, you loosen the grip of your ego and broaden your identity. As soon as your awareness merges with something else, you lose your separateness and become a blend. You are not the distinct personality you were a moment before. You may be speaking as a conscious kitchen table now. You'll eventually feel so connected with life that you'll be able to feel other people's feelings and sensations and think their thoughts—if you want to. In the long run, this teaches you a tremendous amount about collective consciousness and fellowship, which we'll discuss more in chapter 10.

What I Learned from Drowning in Japan

This communion lesson was dramatically demonstrated to me during one of my visits to Japan. I'd been doing in-depth life readings for clients every day without a break and was standing at a crosswalk at lunchtime. There was a crowd of fifty people waiting with me and a group of similar size on the other side. The light turned green and the two groups fluidly passed through each other like two schools of fish. I, being the only blonde in the crowd, drew a subtle kind of attention. No one looked directly at me, and they hardly touched me, but they were curiously feeling into me as though my insides were an open book. After weeks of this kind of invasion, coupled with a deep daily contact with the Japanese subconscious mind— which I had discovered was full of various ideas of self-sacrifice—I was wearing down.

I became feverish and overwhelmed, had trouble breathing, and was on the verge of fainting. I wasn't sure I could maintain my personal space and reality. One night, alone in my hotel room, I was flooded with images of the crowds and all the stories of all the clients I'd seen, and felt I was drowning. A "hand" came up from below and dragged me under. In a vision that lasted only a few seconds, I felt myself drown in a sea of watery energy. I died out of my old American, individualist reality and "came to" in a new one. I was totally submerged, swimming in an ocean of fluid awareness where everything and everyone was interconnected. I've since learned that many Asian cultures consider this to be normal.

I could stretch up out of this sea of feeling without leaving it, like a cresting wave, and take different shapes; I could be me, or a client, or a person on the street, or a tree. When I rose up inside other people, I knew them as if I *were* them. Then I could relax back into the energy ocean. I suddenly understood the Japanese reality from within it. From that day on, I didn't feel crowded because I didn't feel contained. My temperature went back to normal and I felt connected and happy about everyone I saw. They all felt like family, and I could see that in this inner reality, everyone knew about each other and took care to protect each other because one person's pain affected the rest. And that was why the Japanese placed such a high priority on "saving face"—it hurt everyone when any one person was ridiculed or made uncomfortable. This experience changed me totally. It demonstrated a truth about human interconnection that is today beginning to be experienced by everyone, not just Asians, and which is a big part of our future reality.

When you start using senses you've neglected,
your reward is to see the world with completely fresh eyes.

Barbara Sher

Trust Your Inner Perceiver and Follow the Path
of Greatest Resonance

Part of working well with conscious sensitivity is learning to tell the difference between what your soul wants you to notice and things that aren't particularly useful. There is a force inside you—call it the Inner Perceiver, the Revealer, the Holy Spirit; there are many names for it. It's the power of your soul directing your attention to notice things that help you learn your life lessons and express yourself authentically. I *know* my sister, and we grew up in the same house with the same parents, but our recollection of childhood experiences makes me laugh at the vastly different worlds we seemed to have occupied. Our Inner Perceivers were obviously causing us to notice entirely opposite things. She had a problem with our mother's policy of having us eat a small sampling, or "no-thank-you helping," of every type of food on our plate, while I thought it was fair. She grew up to be a PhD nutritionist, and I grew up to eat just about anything.

You can trust your Inner Perceiver and the sequence of things it causes you to notice. The trick is to deepen your involvement with what you notice, asking your Inner Perceiver, "Why am I noticing this person?" "What shall I do with this perception?" Perhaps you take particular notice of a handicapped person as you're waiting at the airport. Why? They may be demonstrating a kind of practical courage you need to apply in your own life. You can also negotiate with your Inner Perceiver. If you're worried about having a car accident, for example, instead of living in a paranoid state, you can ask: "If there is a real danger of an impending accident, please make me aware of it so I can avoid it. If there are other threats around that aren't important to me, don't make me aware of them. I only need the information I can do something about." The key then is to relax and trust your conscious sensitivity and Inner Perceiver.

When navigating via sensitivity and vibration, choose people and opportunities that feel deeply comfortable and that resonate harmoniously with your home frequency. You can distinguish people, opportunities, places, and answers that "make beautiful music with you." Some people or situations might resonate at the same tone as your home frequency, like middle C, and some might be at a higher or lower octave of it, like high C or low C. Others might resonate to middle C but have a different timbre from you. A violin playing middle C sounds different than a tuba playing middle C, for example, but they might sound great together. When you encounter people, ideas, and opportunities that are dissonant with your home frequency, don't get involved—or if you must, let it be peripherally.

> We are perceivers. We are an awareness; we are not objects;
> we have no solidity. We are boundless.
>
> Don Juan/Carlos Castaneda

As you contemplate engaging with new people and experiences, here are a few reality checks. Ask yourself:

1. Is this option on my wavelength? Is it a higher or lower octave of my home frequency? Does it harmonize with me?
2. As I imagine the potential interaction, are there natural flow, cooperation, easy communication, and mutual support and benefit? Do I feel better because of my involvement? Does this option help me be even more sensitive, empathic, and aware, or will I have to guard against shutting down?
3. Will I need to educate, support, convert, struggle with, or do more than my share to have this option be successful? Will I sacrifice myself in any way?
4. What would I learn by contributing to and being affected by my involvement?

It's perfectly fine to say no to tempting offers. The more you choose *not* to engage with dissonant vibrations, even if they're only a little bit "off," the more energy and authentic creativity you'll have. You may have been able to tolerate a job that drained you before, but now it will feel like it's suffocating you. Set your own tone. By inviting others to play with you at a higher frequency, you add greater credibility to the new Intuition Age way of being.

Update Your Reality to Match Your Home Frequency

You've learned to keep your personal vibration attuned to your home frequency by repeatedly choosing to feel the way you love to feel. Now, as you expand to include more of the world, still maintaining your home frequency, you'll notice many situations that seem prehistoric and boring, processes that are snagged and slow, and people who aren't as far along as you are. You may find yourself out of synch and no longer resonating with the way things have always worked or with business partners, spouses, friends, family, and clients. You may feel taxed by the primitive, unconscious, or unscrupulous ways people are doing things.

Two businesswomen, both wise and experienced spiritually, recently shared similar stories with me. One has been in charge of a "green" real estate development project, raising immense sums of money, managing emotionally erratic business partners, and solving unending problems with the city and bordering landowners, while not succumbing to pressure to abandon the project's original spiritual vision for expediency's sake. There have been waves of success, then setbacks. She is currently facing the prospect of a new financial partner who is forcing her out, basically stealing the project from the founders and holders of the vision. In spite of being skilled at staying in her home frequency, she is on the verge of bankruptcy, her body is burned out from too much adrenaline, and she's at that tipping point where she must let go and recenter—yet again.

The second woman went through a similar process with the company she owned. When the market turned sour, her company couldn't pay its bills. Everyone was angry with her; she had gained weight, was depressed, and was talking to a bankruptcy attorney. Then one day, it dawned on her that she might as well stop worrying—after all, what good was it doing? Not long after that, one of her products captured great public interest and sold well nationally and internationally. The tide turned. She had let go and reconnected with her home frequency, and life began to reappear in a new, more charmed way. Her health returned, she lost weight and paid off her debts, and her company was bought by a big-name company. She was given her own division with the freedom to produce products she felt strongly about. Now her intuition was sharper than ever. She recognized the potential in another product and jumped on it. This one became even more successful than the first. Then her new partners engineered a hostile move to take the product and cut her out of future profits.

Both women are learning a similar lesson. Both committed to ambitious endeavors that inspired them spiritually. They had decided, perhaps unconsciously, to expand themselves quickly, and as a means to an end, they chose to partner with people who made money through glamor, one-upmanship, and celebrity. By creating from their high home frequency, they lived through instances in which they experienced magical results. Then, by choosing partners whose vibration was actually much lower than theirs, they were able to see how fear and ego—the "old reality"—could ruin a good thing. Both women now need to reestablish their home frequency and choose new partners and new opportunities that resonate harmoniously with them. They are in a process of *frequency sorting*.

> The voice of our original self is often muffled, overwhelmed,
> even strangled, by the voices of other people's expectations.
> The tongue of the original self is the language of the heart.
>
> Julia Cameron

Fill Your World with Your Home Frequency

You may be at a similar place, where you're experimenting with vibrations and how they affect your life and success. You may have experienced the magical results your home frequency can create and the snags and failures that dissonant relationships and opportunities precipitate. You, too, may now need to entitle yourself to have your home frequency be the basis of everything in your life, to know that your reality can be fluid and joyful in *every area*. To the extent that you haven't accepted harmony as a way of life, you will still attract people who challenge or negate your reality. To the extent that you are partly clear and partly in your old unhealthy feeling habits, your life will contain a mix of fear-based "old reality" thinkers and soul-based "new reality" thinkers.

You may have reached a similar place, as Cameron did in the previous chapter, where you've put more of your energy into situations than was necessary or "put up with" difficult people, where your energy stabilized others and you tried to keep yourself on an even keel so you could do it all again tomorrow. As you sense your expanded capacity and realize you don't have to sacrifice any part of your self-expression for success, you'll realize that you can imagine "striking the tone of your tuning fork" and sending the vibration of your home frequency out through the field around you as the organizing frequency of your world. The more you accept that this is possible and can be real, the more you'll see the people and situations in your life coming into resonance with your home frequency. And then life miraculously improves. You'll soon see that you don't need to lower your own vibration to be in the world—it's just as easy to create a successful reality from your home frequency and have friends and colleagues who understand and support you.

Try This!

Feel into a Power Spot

By using your conscious sensitivity to find places that have power or that resonate to your home frequency, you can maximize your energy level and health, increase your clarity, and even accelerate your growth.

1. Center yourself and calm your body. Begin wherever you are—in your house or yard, at the office or the market, or on a hiking trail. Feel the energy inherent in the spot where you're standing. It's coming directly up from the center of the earth into your feet and body and filling the space around you. Now move slowly, exploring the surrounding space. Let your body find a place it wants to stop. Notice the energy there. What's interesting about it?

2. Let your body move again and stop. Notice the energy. Keep going until you reach the most magnetic, highest-energy spot you can find. Notice the difference between this place and the other places. Breathe in as much of the energy from this power spot as your body wants and send your love down into the earth and into the place.

3. Try this at a restaurant to know where you want to sit or when you're out in nature, arranging furniture, or planting a garden—where do you want to walk? Where does each accessory or plant want to be? Find the

spot in your home that conveys the deepest stillness and use it for meditation. Find an old, special tree that helps you think clearly when you lean on it.

Use All Three Levels of Your Brain
to Refine Your Sensitivity

There are three levels of your brain, as indicated in the following diagram, and each level perceives differently. Moving up and down through your brain is like traversing a scale or ladder of awareness. The upper *neocortex*, with its left and right hemispheres, handles abstract-conceptual perception. The left hemisphere governs analytical thought and language, and the right hemisphere pertains to pattern recognition, intuition, and creativity. The *midbrain* is the level of sensory perception, and it helps you feel connected to the world through similarity and affection. The lower *reptile brain* is survival oriented and perceives through instinctive attraction-repulsion and fight-or-flight responses. It is responsible for motivation and direct experience. Since body data rise up the spine, the reptile brain is the earliest point where information becomes conscious. It then takes on extra dimension in the midbrain as the senses flesh it out and moves on to the upper brain where it takes on meaning and language and becomes part of a larger pattern of knowing. If you keep the image of a ladder in mind,

The Three Levels of Your Brain

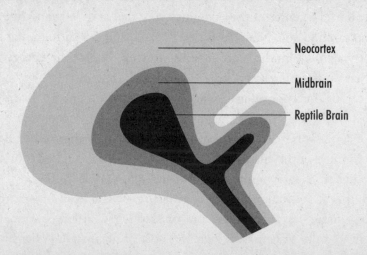

Neocortex

Midbrain

Reptile Brain

you can move fluidly up and down the rungs, shifting easily from body sensitivity to intuition to pattern recognition to meaning.

To emphasize sensitivity, move down the ladder toward body knowing. To emphasize abstract perception, move up. If you've been living in your left hemisphere all day, you may be saturated with words and beta waves, and that can make you feel stressed and jittery. If you imagine moving to the right hemisphere and down to the midbrain—where you experience intuition, the senses, similarities, and affection—you can shift to the slower alpha and theta waves and alleviate tension. (Delta waves, remember, occur mainly during deep sleep.) If you drop further to the reptile brain, where you experience instinct and feel a direct connection to the environment, you can ground out any last remnants of stress. By coming "down" in your brain, you enter your body more deeply and calm yourself. Once you're calm and feeling steady, it's easy to feel your home frequency existing in all the levels of your brain and body.

Kinds of Sensitivity in the Three Levels of Your Brain

Early Recognition **REPTILE BRAIN**	Senses and Feelings **MIDBRAIN**	Most Refined Sensitivity **NEOCORTEX**
subtle "vibes" heebie-jeebies/ butterflies gut instinct attraction/repulsion expansion/contraction resonance/dissonance	smell/inner sense of smell taste/inner sense of taste touch/ clairsentience hearing/clairaudience vision/clairvoyance empathy/communion	rational meaning flashes of understanding sudden grasping of patterns mysticism feeling nonphysical beings unified field awareness collective consciousness

To activate very subtle sensitivity, it helps to close your eyes and drop below your sense of sight—since vision and the inner sense of sight, *clairvoyance*, are closely connected to your neocortex—and shift you into language, meaning, and concepts almost instantly. To develop sensitivity, you want to stay merged with your body. Try listening to sounds—inside and around you, near and far, soft and loud. Then make a subtle shift so you can "hear" the sounds below the audible sounds. You might hear messages from your organs, a friend's voice, or the plants. Be careful not to go off on tangents, as *clairaudience* ties to your inner voice, and that can lead to too much self-talk, which takes you back into your upper brain.

Remedy that by moving below sound into tactile awareness. With your eyes closed, sense what's inside and around you by feeling temperature and textures. You might feel that your head is hotter than your feet, there are temperature differences in the air, or a fabric is nubby. Shift to your inner sense of touch—*clairsentience*—and receive impressions. Does your energy have a texture? Is the energy in the room you're occupying stagnant, electrical, or nurturing? Now try this: combine your hearing and tactile senses. See if you can feel the physical impression that the ticking clock or ringing phone makes on your body. With eyes still closed, notice what you smell. The quieter you are, the more you'll notice subtle odors, like the difference between a plain piece of cold paper and one that's just come out of a warm printer. Shifting to your inner sense of smell, you might receive impressions about what the health of your body smells like or what a potential solution to a problem smells like. You can do this with taste as well, sensing that a situation is sweet or bitter, for example.

I am convinced that there are universal currents of Divine Thought vibrating the ether everywhere and that any who can feel these vibrations is inspired.

Richard Wagner

By attuning to subtle perceptions and moving down the ladder toward your instinctive senses (see the Scales of Everyday Vibrations chart in chapter 2), you'll develop conscious sensitivity. Your sensitivity brings information about other bodies, objects, places, events, trends, timelines, and processes—even from the near future. You may suddenly understand the nature of emotional wounds and healing, motivations, the complex patterns of thought behind behaviors and results, and eventually, enlightened emotional states, like ecstasy, joy, and bliss, as well as the goings-on in higher dimensions of awareness.

Try This!

Feel into Financial Investments

If you're contemplating buying or selling stocks or mutual funds, or making other investments, make a list of your options on paper and sit quietly, looking at it. Focus on the first item: should you sell this mutual fund or keep it? Feel into it with your expanded tactile sense. What impressions do you get? Is it a high vibration, an erratic one, or does it feel solid and real over the long term? Do you get any twinges, sinking feelings, spiky sensations, or perhaps a feeling of emptiness? Does the option make you feel happy, dull, scared, tight, enthusiastic, or unconscious? You might even merge all the way into the mutual fund, become it, and speak as the fund, giving yourself a message. Note your impressions next to the item. Move to the next and the next. When you're finished, you might do some research, see what other people are saying, and check whether there is some evidence for your insights. Then take action.

The Sooner You Know, the Better Your Sensitivity Serves You

How early do you recognize nonverbal information coming to you via vibration? The sooner you register an impression and decipher it, the more consciously sensitive you'll become. You might start by being conscious of the things you *do* notice easily and early. Like most people, you're most sensitive to what you think is important or fascinating. If your fears and worries are predominant, you may be hyperalert to threats to your physical well-being, financial security, or marriage. On the other hand, if you've made a commitment to your growth, you may notice positive things that help you expand, like a phrase that opens your understanding, an urge to get some sunshine, or a new course of study that begs further exploration.

Next, you might rev up your curiosity to see what your body is picking up on during the day and on what subtle information or impressions you're basing many of your quick, navigational decisions. Why did you suddenly stop working and decide to run a few errands? Why did you return one phone call before another? Why did you speak sharply to a person to whom you're usually kind? Finally, see if you can stretch your sensitivity to become aware of hidden forces or variables that are about to affect you or your loved ones. By noticing what you're noticing, you'll streamline the information-to-appropriate-action circuit.

When your body picks up nonverbal information from the environment as vibration, that vibration will intensify if you don't consciously recognize it and act on it. There are times when you may be temporarily insensitive because you're in your left brain too much, are in one kind of focus too long, or are out of your body. But your soul is talking to you, and if the information is important, it won't stop pushing until you get the message. Let's say you become tense at work and are in a bad mood by the time you get home. You may have glossed over the early signals that were telling you that you don't do well with your overwhelming work load or that being under fluorescent lights too long isn't healthy. Since you didn't get the message right away and take action to alleviate your body's concerns, pressure built until you're now close to having a migraine.

If you don't register the first tiny waves of expansion and contraction, the information becomes more "high-pitched" and insistent. Your body whispers first, then clears its throat, knocks at the door, bangs on the door wildly, then turns on the red flashing light and siren. Subtle contractions build to tension, to pain, to chronic pain, to illness, to paralysis; trauma and accidents eventually occur. Underneath every pain and trauma is an unheard message from your soul and body. So picking up subtle information early, through conscious sensitivity, can save you a lot of pain and worry!

Try This!

What Are You Most and Least Sensitive To?

Bringing the following categories of experiences into your awareness can help you be more consciously sensitive to what you're feeling.

- List the things that "push your buttons," things that can get a rise out of you.
- List things that scare you at a deep level and produce anxiety.
- List things you want to learn that you might find embodied in other people.
- List things that other people notice or feel that you seem insensitive to.

- List ways you associate your five senses with feelings of pleasure and beauty.
- Write about how early you notice tension, pain, or minor illness in your body.
- What are you sensitive to in the environment (light, dampness, color, foods, ugliness, temperature, noise, altitude, etc.)?
- Write about how you could notice nonverbal information from the environment at an earlier stage.
- What are your most dominant senses? Which ones would you like to develop more?

The Subtle Vibration of Your Truth and Anxiety Signals

You are highly sensitive to truth and lies, safety and danger. When information registers in your reptile brain, it takes two forms: *yes* and *no*. You'll recognize these cues through instinctive feelings of expansion or contraction, of energy moving or not moving. Be careful not to dismiss these sensations as mere idiosyncrasies in your body—they are your very own conscious-sensitivity discernment system! When a choice or action is appropriate, safe, true, and on purpose for you, you'll feel expanding energy: you may sense energy rising, becoming active or bouncy, or perhaps you get "hot" for an idea, feel "light-headed," or flushed with enthusiasm and "butterflies." These responses are your *truth signals*, and you can feel them in various parts of your body. Some people experience a warm, spreading sensation across their chest. Others feel energy bubbling up from below the diaphragm into their chest, throat, and eyes, which might brim with tears. Some feel blood rushing to their neck and face, making them blush, or energy moving up the spine or down their arms, giving them prickly chills. Still, others describe "clicks and clunks," as if something comes into alignment and drops into place.

When an option or action is unsafe, inappropriate, false, or off purpose for you—you'll experience an *anxiety signal*. You may feel energy drop, recoil, darken, or tighten. Maybe you'll feel cool or cold or get a sinking feeling in the pit of your stomach. You may feel repulsed, become leaden, or "turn to stone." Instead of blushing, you may blanch or feel gray or blue. You may feel pain or have a literal "hair-raising" experience. Common anxiety signals are: stomachache or nausea, a "pain in the neck," tightness in the chest, headaches, or a knot or fistlike feeling in the solar plexus.

A man in one of my classes did an exercise to use his sense of smell combined with truth and anxiety signals to solve a problem. He had three possibilities. When he imagined each solution and what it might smell like, one smelled like grilled steak, one like dirt, and one like a cut orange. Which was he most drawn to? He said his body liked them all. When I asked him to feel the emotional states or the kind of energy that each might provide, the choice instantly swung in favor of the orange because he liked the high enthusiastic feeling of movement it gave him. The other two felt steady but boring.

If you walk into a house and feel a chill that isn't related to physical temperature, what might that mean? Perhaps someone recently died in the house, it's haunted, the occupants had a bitter argument, you're about to experience disappointment or bad luck associated with your stay there, or it reminds you of a past negative experience that you need to understand and clear. If you meet a potential business partner and your heart starts pounding rapidly, does it mean that you've met your soul mate or that you should double-check with your home frequency to see if the person is one of those danger-in-disguise types? If you need to make a phone call but keep putting it off, does it mean that there's something about your relationship with the person you need to clarify or that you're missing a piece of data that might change what you're going to say? These are the kinds of things to pay attention to and to practice decoding.

> You are standing on the seashore and the waves wash up an old hat,
> an old box, a shoe, a dead fish, and there they lie on the shore.
> You say, "Chance, nonsense!" The Chinese mind asks,
> "What does it mean that these things are together?"
>
> Carl Jung

You Can Apply Sensitivity in Business

David, a corporate recruiter and management consultant in the San Francisco area, says that when he first expanded his consulting practice to headhunting, people told him he'd have to make at least fifty cold calls a day. But David was experimenting with a philosophy of how everything in life was inter-related and cooperative, so when he started his new practice, he faced his desk toward the east so he could feel the bulk of the United States. Each morning he'd sit at his desk, close his eyes, and connect with the whole country. He'd visualize and feel thousands of points of light peppering the land, all the way to the eastern seaboard. Each point of light was a person, connected by a thread of light back to him at his desk. Next, he reviewed the job listings and got a sense of the kinds of people he needed. He then happily sent the requests out along the lines of light, as though he were striking a dinner bell, calling people to come. He saw certain points of light swell and grow larger.

Then he let go of the image, let himself feel content, and turned his attention to his "To call" list. The name that jumped off the page at him first was the one he called. In a short time, David was having phenomenal success. His colleagues didn't understand why he was so charmed. Instead of fifty calls a day, he made ten—but they were the right ten. Most of the positions he filled came from people calling out of the blue, or from referrals. Clients said they enjoyed his ease and how things magically materialized around him.

A number of my clients who are successful entrepreneurs and CEOs tell similar stories of using unorthodox methods based on sensitivity. Peter, who had risen to a high management position in an investment firm, told me that when he took charge of a huge department, he was given a glass-enclosed office, which he couldn't stand to be in. It cut him off from others, and the only way he knew what strategies to create and implement was to walk around and talk to people. He often just chatted about their families or vacation plans. He realized that his body picked up information that wasn't verbalized; this knowledge melded together into an overview, and when he had to go back in his "cell," plans emerged effortlessly. He felt comfortable in his role as long as he could "stroll."

Your Body Is a Barometer of What's Happening Around You

I can't count the number of times this has happened: I tune in to my body to see what I'm experiencing, looking to recap a movement of conscious-ness and energy so I can write about it in my newsletter or share it in a class. I might feel frustration and panic or waves of inexplicable relief.

I may realize I've had a breakthrough into a new level of awareness. When I share the insights, I get a flood of responses—way more than statistics would probably predict—saying, "The same thing happened to me!"

An entrepreneur I knew walked through her company's offices one day, sat down at her desk, and felt something vague bothering her. She sat with it a while and let it surface. What came to her wasn't pleasant news, but it was useful. She realized her employees were *bored*. How awful! She'd created a company and working conditions that weren't alive or interesting enough to motivate the best in people. What was she to do? She sat with it a little longer and had another revelation: *she herself was bored*! The repercussions of this stunned her—that she and her company and her employees were so intimately connected that a shift in her own attitude would have a ripple effect throughout the entire place. At times like this you wonder, "Which came first, the chicken or the egg?" Was her boredom contagious at a telepathic, vibrational level to the employees, or were the employees radiating an energy that affected her? Or was the whole system responding at the same time to changing energy conditions environmentally? The answer is D, all of the above.

> Take care of your body with steadfast fidelity. The soul must see through these eyes alone, and if they are dim, the whole world is clouded.
>
> Johann Wolfgang von Goethe

Philip was married for a short time to a woman who had an anger problem and regularly vented at him for fairly innocuous reasons. In the beginning of their relationship, her fits of rage would take him by surprise and he could feel his solar plexus and chest contract suddenly, as though from a real physical blow. He tried to let it pass, let the intensity roll off his back, and not react in kind, but his body couldn't stop recoiling painfully. After a year of this, he noticed his body began to be anxious and jumpy just prior to one of her violent outbursts. It was almost as if his body could smell the "ozone" building up in the environment around him or feel some sort of extremely subtle tension. His body became a reliable barometer of the changing emotional weather in his relationship, accurately predicting, often within hours, the onset of his wife's anger cycle.

Your Body Can Know Things in Other Times and Spaces

These kinds of experiences are not at all uncommon. In response to a crisis, the death of a loved one, or even a fortunate windfall, people often say, "I knew it. I felt it coming." Your body registers information about your immediate environment and relationships, or about an event that's about to occur, because waves of vibration are constantly bringing subtle data. Perhaps the "quantum entities" in a pattern of knowledge of a distant or future event dissolve into waves and reappear in our bodies as impressions as they become particles again—bypassing time and space altogether.

In the unified field, distance doesn't matter; time doesn't factor in. Knowing is immediate and everywhere at once. Think of your body as the central focusing lens of the universe *for you.* My body is the central focusing lens of the universe *for me.* Whatever is happening in the larger field that pertains to your soul's concerns filters down through your personal field and focuses into your body-lens where it can be consciously recognized as though you're gazing into a crystal ball. This process intensifies and accelerates a hundredfold when you intentionally look for insight or guidance.

You Can Receive Impressions Easily
from Resonant Realities

If I overlap with you by sharing similar life lessons and growth sequences, our realities will be resonant. It will be easy to know *you* by the thoughts *I* am thinking, the emotions and urges *you're* feeling by what *I'm* feeling. We're sharing a common energy field. In much the same way you'll find that people who share a physical environment often have similar lessons. People who work in a company that expects many hours of overtime are all, in some way, looking at an issue of self-sacrifice. People who live in a country controlled by a dictator are all, in some way, learning to find their inner authority and authenticity. Do you want to understand the political trends in your country? Look to what's happening in your own energy, emotions, and thoughts. What are your issues? What is your leading edge of growth?

Imagined realities are also resonant or dissonant, and your body will tell you accurately whether a potential future is viable or not. Do you want to know how it might be to work with a new business partner or to invest in a piece of real estate? Your body will have an immediate reaction to the imagined future reality. Do you want to know how a process will unfold? Imagine it and your body can "read" the blueprint of the overall flow of events, expanding and contracting along the way as snags and breakthroughs are likely to occur.

Try This!

Feel into Dissonant or Resonant Vibrations

1. List three acquaintances. Imagine each person's body radiating their personal vibration toward your body. As it meets yours and begins to pass through you, does it synchronize easily with yours? Or is it rough or "off" somehow? Imagine your personal vibration, focused at your home frequency, radiating toward and through them. Does it resonate easily? Does the person adapt to harmonize with you? Or is there even the slightest bit of dissonance? Now try the same thing with a best friend and feel the difference.

2. List three places you'd like to go on vacation. Imagine your body in each experience. Read your body's signals and vibration for the amount of resonance or dissonance it has with each place.

3. List three tasks you need to do. Imagine your body doing each one. Read your body for the task that holds the highest resonance because it matches your home frequency, or because it is on the universe's "urgent" list, or because it offers something you need first. Rate them in order of your body's priority.

Try This!

Describe Subtle Sensations

Write about what the following subtle sensations feel like or how and where you recognize these things in your body:

- How do you feel when something is trying to make itself known?
- How do you feel when you override an intuition?

- How do you feel when you shift from a personal view to a collective worldview?
- How do you feel when something's about to go wrong?
- How do you feel when something's about to go very right?
- How do you feel when you're picking up information from another person's body?
- How do you feel when you recognize something as a "sign" or symbol of something?
- How do you feel when you're receiving a wave of energy that contains data?

Here Are Some Tips for Becoming More Consciously Sensitive

When working with energy and awareness, knowing a few more key *frequency principles* can help you decipher the data encoded in waves and vibrations.

1. **There is a right use of will when feeling into.** When you expand your attention through the field around you to know with your sensitivity, don't "push" your way out. Simply remain open, soft, curious, and expansive. Let yourself receive what wants to come. Let yourself notice what your Inner Perceiver wants to show you. Your job is to expand and be impressed. If you look into someone's eyes, for example, don't force your way in or you won't see anything. Let your eyes relax and be receptive, and you'll receive a continuous stream of information.

2. **There's a difference between feeling into by "going over to" versus including.** If you hold a worldview based on separation, you'll perceive that you are "going over to" feel another person, idea, event, or process—and you'll have a linear image of a "connection." You're crossing a gap, and this takes willpower, as well as promoting the idea that you have to leave your own center to find what you want. When you do this, you must recenter after each foray. If you catch yourself knowing this way, make sure to come back to yourself and say, "What do *I* think about this? Does this work for *me*? What is *my* version of this?" If you don't recenter, you may go off on a tangent, get stuck living in the other person's reality, or feel drained.

Remind yourself that you're always in the center of your personal field and you can simply expand that field of awareness to include the thing you want to understand. You never have to leave "home" and your home frequency. Your field will do the sensing for you. This way, it's easier to merge with what you want to understand, feel it as an aspect of yourself, keep your heart open, and know from conscious communion.

3. **It's a mistake to think you alone feel something.** When you share responsibility with everyone for knowing things, answers (and questions) emerge from the field magically, just as you need them. You'll find that everyone is a messenger of the one great unified Self. All bodies, even those of animals, birds, insects, and plants, can relay vibrational information.

4. **You receive what you need to know—to do what you need to do—in the present moment. There are good reasons you may not feel too far into the future.** You notice what's meaningful, and things are meaningful because you're learning a lesson related to perceptions. You receive information as you need it. As soon as you've used the information and integrated the experience it's meant to provide, you magnetize the next idea and experience. You may not be able to know too far ahead, or your intuitions about the future may prove inaccurate because you've been ignoring an important perception right under your nose. By integrating and using perceptions that may be stacked up like planes waiting to land, you clear a wider view.

5. **Sensitivity is empowered by trust and validation.** If you make an agreement with your soul and body to trust the subtle information that comes via vibration, to notice cues that indicate a message is waiting to be communicated, to decipher nonverbal messages to find meaning, to use the information to be a better person, and to validate the whole process regularly, you'll optimize your conscious sensitivity skills. When your body gives you a message via sensitivity, thank it out loud, and hug, pat, or stroke it lovingly. Bodies love sensory input!

6. **Emotions are exaggerated sensitivity signals.** Whether your emotions are expansive and feel good or contractive and feel bad, you've probably been ignoring earlier, more subtle sensitivity clues, and the energy has built in intensity to attract your attention. Your emotions bring information about direction, what your soul wants, and what life lessons you need to pay attention to.

Qualities of a Consciously Sensitive Person

A consciously sensitive person:
- has a strong sense of self and a malleable personality;
- is comfortable being a Me and a We; a particle, a wave, and a field of awareness;
- enters, merges with, and becomes what he knows;
- is deeply caring and neutral and uses the empathic state as the Teacher;
- perceives the outer impersonal reality as an inner personal one;
- trusts that subjective knowing parallels objective reality;
- is aware that the physical body is conscious and alert;
- uses her personal vibration to catalyze what she needs;
- works with universal energy to create and heal;
- heals by reestablishing oneness and love, eliminating doubt;
- heals others by healing herself, knowing that others are inside her;
- feels "just right" movement and choice, and his choices feel "deeply comfortable";
- suffers by being out of harmony with life and feeling people who feel separate and isolated.

Just to Recap . . .

You can engage with experiences and navigate through life by using your sensitivity to consciously feel into the world. To do this, you inch your way out through your environment, including and merging with the people, objects, situations, processes, and events you encounter. Instead of connecting with them, you enter a state of conscious communion and know them as aspects of yourself, from the inside. This fosters caring and compassion. It's important now that you imagine "striking the tone of your tuning fork" and sending your home frequency out through your personal field as the organizing frequency of your whole reality. That means you may need to do some sorting and housecleaning to reaffirm the vibration you choose to have as the basis of your relationships, work, and self-expression.

Each of us can manifest the properties of a field of consciousness
that transcends space, time, and linear causality.

Stanislav Grof

Your clarity increases the more you unplug from dissonant vibrations and
self-sacrifice. You know what works for you by sensing what's in harmony
with your personal vibration, whether it's a lower or higher octave or a different timbre of your home frequency. You interpret the subtle data encoded
in waves and vibrations based on how your body responds: either through a
truth signal (expansion) or an anxiety signal (contraction), or via pleasing or
displeasing sensory information. The earlier you pick up information the less
it will build to stressful levels. Your body is the barometer of what's happening around you and even of the near future. It's natural to directly share
knowing through vibration with others whose frequency is similar to yours.

Home Frequency Message

As I explain on page xxi in *To the Reader*, I've included these pieces of inspired
writing at the end of each chapter as a way for you to shift from your normal,
speedier reading mind to a deeper kind of direct experience. Through these
messages, you can intentionally change your personal vibration.

The following message is meant to transport you into a way of knowing the
world that's close to the way you'll experience life in the Intuition Age. To
move into the *home frequency message*, just downshift to a slower, less hurried
pace. Take a slow breath in, then out, and be as calm and still as possible. Let
your mind be soft and receptive. Open your intuition and prepare to *feel into*
the language. See if you can experience the deeper realities and feeling states
that come alive *as you read*.

Your experience may take on greater dimension in direct proportion to the
amount of attention you invest in the phrases. Focus on the words a few at a
time, pause at the punctuation marks, and "be with" the intelligence delivering
the message—live and right now—to you. You might speak the words aloud,
or close your eyes and have someone else read them to you and see what effect
they have on you.

ATTUNE TO THE TRUEST FEELING

You are so much more unlimited and privileged and well served than you realize! All things are at your fingertips, awaiting your permission to occur. All knowledge waits only for your curiosity to draw it forth. The living field of presence in which you reside, of which you are made, is sensitive to the desire of a single atom; it responds and shifts and opens toward your every tiny need. Flowing with its willingness, you can easily feel the naturalness and pleasure of service. Yes, you are discovering the benefits of conscious, refined sensitivity and harmonious resonance—how it makes for living well and accessing the ascending spiral of wisdom. But where, really, does conscious communion lead? It leads you to an experience of the Us, through shared knowing, shared feeling, and shared motivation, into the truth of your Belonging. Your communion with your brother-and-sister forms of life shows you the unselfishness of the world. Look around at the simple details and feel into the motives of objects. Why do they exist? There is but one real feeling: Generosity.

The café welcomes you, and you notice and feel the red toenail polish flashing on the woman's happy toes, her sandals showing them off, and the man's cell phone, living to connect him to others, his shoulder and neck bent toward each other to hold it in place so his hands can reach his money, that goes wherever he wants to send it, willingly, never having much of a home. The baby's fat legs like to make you smile, dangling from his father's kindly forearms. The round granite slab, polished into a tabletop, supports many arms, cups, books, and scones, and patiently lived as a mountain until it was sliced away from its family to come to your town, to show its green and black flecks so beautifully and continually. You can feel the pen's love of making marks, the hand with fingers' joy in knowing how to press and shape letterforms, in conveying motives from the brain, through the neck, shoulder, arm, and wrist onto the waiting paper that came from a tree that grew in Oregon and died after a not-long-enough life so you could bring your thoughts, these ephemeral things, into this world that so loves itself that all it knows to do is Give.

7

Mastering Relationship Resonance

The profound feeling of connection and belonging that is evoked
in the experience of "home," and in the presence of someone who comes
from our home or with whom we have made a home, reflects a kind
of matching, a pervasive resonance between what is inside us with what
is outside us, between the past and the present, between what
we were, what we are, and what we long to be.

Stephen A. Mitchell

If you've been reading *Frequency* in a continuous way, you'll notice we've moved through a few stages of the transformation process. We first got the lay of the land in terms of frequencies and how your awareness is accelerating, then looked at what might block or distort your sensitivity and how to clear it. After that, we focused on finding your home frequency and worked to develop conscious sensitivity skills. Now we'll move into another phase: applying sensitivity and frequency principles to some important areas of life. The first of these is relationships, for it seems everything we do is based on relating—even meditating. In addition to people and animals, we have relationships with everything from our money and car to our body and the Divine.

Having great personal relationships—those we might even categorize as soul mates or spiritual family—is our most idealistic dream. What's more satisfying than being on the same wavelength with someone? Or working together harmoniously, anticipating each other's needs? It's a primal joy to assist in someone's healing or to offer help when someone needs it. But everyday relationships can also make us miserable, bringing our most challenging trials and traumas. What to do? Relationships, just like everything else, are vibrational, so knowing how to work with frequency principles can turn troubled interactions into real gifts of spirit.

Who's Populating Your World?

Take a look around. What sort of people do you have as friends, business associates, and intimates? If you're lucky, they're trustworthy, affectionate, and thoughtful. They have time for you and want the best for you. They meet you halfway, coming as far forward into engagement as you do, and they're adaptable, communicative, productive, and contributory. On the other hand, you may be surrounded by people who are wounded victims, leeches, blaming punishers, withholding noncommunicators, or passive-aggressive dominators.

There's a reason your world contains the people it does. All relationships, even those that seem difficult, are gifts from your soul to help you find blocks, talents, and new directions that you might not discover alone. Some people might share interests and affirm your talents, helping you develop confidence. Some are catalysts to help you learn your life lessons. Some introduce you to new ideas and potentials. Still, other people may be with you simply for joy's sake because your souls like being together. Troublesome relationships exist to help free you from unhealthy feeling habits.

> Love is our true destiny. We do not find the meaning of life
> by ourselves alone—we find it with another.
>
> Thomas Merton

Relationships Are a Path to Transformation

Relationships speed your transformation process because they help you see yourself faster and more clearly—both your positive traits and your blockages. It's hard to avoid the emergence of core issues and growth in the energetically amplified environment that results from two people combining their vibrating fields. In this intensified resonance, you can clear unhealthy feeling habits more quickly and discover new aspects of identity, which, in turn, expands you. You soon see yourself as a much larger, more complex sort of being. The effects of love and fear are immediately telegraphed back to you in relationships, so it's easy to quickly test what works to bring more soul into your life and what doesn't. And relationships propel you beyond the idea that only *you* have certain problems, that only *you* have a certain potential, which helps develop empathy. You see that everything you experience can be experienced by everyone, and vice versa—and that's empowering.

If our egos had their way, we'd stay stuck in the limited view of "I am me, and you are you, and never the twain shall meet." But you're transforming now, and that means you're capable of knowing yourself as a vibrating being made of energy, and knowing how you and other people are fields of awareness that interpenetrate each other. You now can really understand how we all share everything, know about each other, support each other's existence, and evolve in tandem, never alone. Though it looks complex, you can also now feel how elegantly simple our relatedness is: at the core, all our home frequencies eventually merge into one home frequency. We all vibrate at the universal frequency of love.

When you feel into a relationship you can sense how you belong to a single *relationship field* that contains both people's combined histories, talents, and potentials. This field becomes a sort of guide to both people, and once you can merge with it, you're able to include the relationship's wisdom and natural characteristics, its life lessons, and its destiny as part of your own life. So in addition to knowing yourself as an individual, you're on your way to knowing yourself as the consciousness of the relationship; it's dawning on you that you're both an individual *and* a collective self. Eventually, as you learn to include more people in your pool of relationships, you traverse the scales of identity, feeling yourself as an individual, a relationship, a family, an organization, a nation, humanity, and on into the higher nonphysical dimensions where everyone joins in a unified state and the Self becomes pure Awareness.

The People in Your Life Frequency-Match with You

Whether your relationships are flowing with perfect synchronization or are emotionally upsetting, it's always because the other person has a frequency pattern similar to yours. Those who populate your world *frequency-match* with you. If someone has shown up in your life and if your Inner Perceiver is noticing them, you share a common personal vibration. If they occur in your world, you occur in theirs. You mutually create each other for a shared purpose. They aren't identical to you, of course, but you share similar frequencies. Once you come together, your personal fields merge, and a relationship field comes into existence. That relationship field then sources your experience as a couple, bringing you shared resources, like energy, wisdom from past experiences, and access to new talents. Because of this, it's possible to know things the other person knows by osmosis. Interestingly, people with highly dissimilar vibratory patterns and little in common do not occur in each other's realities.

> We do not attract what we want, but what we are.
>
> James Lane Allen

This matching process can be confusing because (1) it can be conscious or unconscious, and (2) partners often act out opposite sides of the same issue and don't see the underlying issue they have in common. It's great to realize you're similar to people you admire, but you may not want to admit that you could be anything like the people who irritate or threaten you. In any relationship, in each moment, the voice of the relationship field addresses one theme in both people. If one partner needs to travel for a month for work and the other protests about being abandoned, the truth is that both people need independence and space. If partners disagree, each must realize their partner is addressing concerns relevant to both. It's not possible to tell your partner things that pertain only to them; every observation about the other is an observation about both of you. Finding the messages coming from the relationship field and recognizing the shared issues in the partnership, whether positive or negative, are big parts of using relationships for transformation.

Discover How and Why You Connect with Others

For relationships to evolve to a higher level, and to value each other as we must, we need to understand the great love that underlies the reasons we connect and the energy dynamics of how we connect. The popular law of attraction says, "Like attracts like." Then there's the old adage that says, "Opposites attract." Let's be more precise about what really happens energetically to cause relationships. You establish the frequency of your *personal field*, the energy that fills the space around you; it is the radiance of your personal vibration. It can be a low vibration or the high vibration of your home frequency. Either way, people of a similar frequency simply arise or appear out of your field and dissolve back into it, in response to mutual need or curiosity. Of course, you appear in their field the same way. We'll explore the deeper dynamics of "attraction" in greater detail in chapter 9, but for now let's think of attraction as a kind of commonality of frequency.

We don't *really* attract other people the way a magnet attracts iron filings. When souls have natural resonance, they simply appear to each other. When the resonance changes, they disappear. Relationships are always mutually created and dissolved. You may intentionally set out to attract someone, make a wish list, and apply various strategies, but the likelihood is that you are thinking of doing it because your soul wants it, and it's already in progress in the inner realms. You can rest assured that if your soul wants someone to appear, it's going to happen. If it's time to meet your life partner or have an increase in clients for your business, the souls will arrange it. And yes, clearing unhealthy feeling habits, acknowledging your desires, and applying an aligned mental-emotional focus can streamline the materialization process. But the reverse is also true: if your soul doesn't need a connection, no amount of positive affirmations will make it happen.

There are a hundred paths through the world that are easier than loving.
But, who wants easier?

Mary Oliver

If you're *strongly* attracted or excited by someone—be it a lover, teacher, or new friend—it's because you both *really* want to clear blockages, and you both *really* want to validate similarities. We think of those who frequency-match our good qualities and goals as "like attracting like." Those who frequency-match our suppressed negative vibrations, or who act out the other half of a polarity we're stuck in, we think of as "opposites attracting."

It's often the case that strong, sexual "animal magnetism" in the early stages of a romance is caused by adrenaline—the partners actually are scaring each other because each unconsciously represents the person who hurt the other in the past.

It used to be that a hyperactive person could attract their opposite solid-as-a-rock person and each would do their half of the polarity, balancing the other. Today we must both be internally balanced—both capable of being active *and* calm—in order to transform our relationships. Now, if you're attracted by a lover's dynamism, you're probably ready to expand your own self-expression. If you're turned on by a leader's generosity to the underprivileged, you may be healing a wound of neglect from your past. In abusive relationships, the abuser and abused are both victims and both dominators. The souls are dramatizing a pattern to make it conscious so they can clear it and reopen their hearts.

Try This!
Ready Yourself for the Ones You Want

1. If you've been thinking about finding new people of a higher vibration or if you're ready for a new life partner, a new business partner, or more clients, it's a sign that your soul is beginning to orchestrate a new phase of growth. If you're thinking about it, it's in the early stages of becoming real. You don't have to make it happen—just relax and allow it to unfold.

2. Feel into the various qualities of the prospective relationship(s) you want. How does the energy flow? How does the mutual sensitivity operate? How honest and easy is the communication process? How happy are the two bodies together? Imagine it in great *tactile* and sensory detail. Be clear that the feeling tone or vibration of the relationship is actually *your* new vibratory level. It's what you're establishing as the unifying frequency of your most updated personal field. Whoever the person or people may be, they overlap with this frequency and are ready for it to be their new reality, too. Relax and invite them to show up. Know it is becoming real. Wait happily.

3. Now actively radiate the frequency of your new vibratory level. You might imagine there's a lighthouse projecting up above you and its bea-

con rotates nonstop, sending your frequency everywhere, day and night. It's there to help reassure *you* that people will find you quickly, and once you set it in motion, it keeps beaming and rippling the energy outward, maintaining the clear vibration of your field.

When Your Relationships Resonate to Fear . . .

Subconscious fear can resonate through your field and give rise to fear-based people just as easily as your home frequency can materialize loving soul mates and soul friends. When people share misperceptions and unhealthy feeling habits, they work out the error of their ways by connecting with others who have the same underlying contractions. You aren't being punished by having troublesome people in your life; you're being given an opportunity to transform yourself rapidly.

There are many ways in which relationships can be out of harmony; fearful egos can turn the interaction sour, produce pain, and shake the pair apart. If both people are contracted when they're alone, the pattern amplifies to a level where it becomes tangible when they get together. Then it's common for both people to resist each other, projecting what they don't want to see onto the other, resulting in blame, punishment, and rejection. If you're involved in troublesome relationships, it could be caused by any of the following reasons.

1. **Both people are subconsciously adhering to the nonverbal "sonar" feeling patterns established in childhood concerning the way their parents acted with each other and them, taking those patterns to be instructions for all relationships.** You're on automatic, reacting without noticing what's true in present time. Your soul brings people who have the same underlying sonar pattern, who alternate playing the victim or dominator roles with *you*. Your mother was mean to your father, and you became his little sweetheart. Now you easily take men away from other women, or they take men from you. Or your mother suffocated you with needy love because your father wasn't present for her. Now you can't commit to women, and if you do, they dominate you or can't commit to *you*. Or one of your parents died or left when you were small, and now you get angry at your partners whenever you sense them needing their own space and you leave *them*, or they abandon you.

2. **Both people are fixated on an emotional or mental position that's interfering with full soul expression.** Your soul brings someone who duplicates or challenges your views in a way that helps you see them. You developed great expertise in science and now use it to explain every happening in your world, and you meet someone who thinks science is limited, that being artistic and in the flow is best. Or you turned your home and yard into a beautiful sanctuary from the world's chaos; your new neighbors next door make lots of noise, smoke incessantly on their patio, and park junker cars in front of your property. Or you nearly drowned in a boating accident as a child, and your soul brings you a spouse who loves the seashore, sailing, and swimming.

3. **One or both people have** *karma*—**unfinished business from the past.** You did something in the distant past that you want to make amends for and understand now. You find a relationship that lets you unconsciously reenact a situation in which you were closed-hearted before but now want to be generous. Or you allow another person to make amends to you for their past misdeed. In another life, or early in this life, you stole from your boss; now your father leaves you out of his will. In another time, you died young, leaving your spouse alone with small children. Now, instead of leaving your difficult mate, you stay and deal with him or her and your problematic stepchildren longer than any "sane" person would.

A karmic relationship is automatic and hypnotic at first, then often feels like you're in a tunnel; you can't get out until you emerge at the other end. You chose it, want to do it, but it's not that much fun. When it's over, you shake yourself and say, "What just happened to me?"

> The meeting of two personalities is like the contact of two chemical substances: if there is any reaction, both are transformed.
>
> Carl Jung

When Your Relationships Change and End . . .

Just as two similarly vibrating personal fields create relationships, relationships change and often end when the vibration of one person's field shifts. If the vibration of your field changes, as it might after a spiritual breakthrough, certain people cannot occur in your field anymore, unless they have a matching breakthrough, and they will likely disappear. You may write it off with comments like, "They're boring," or "For some reason, they don't like me now." Friends drop away, and new ones appear out of the blue. When a karmic period is finished, the vibration of one person's field also changes instantly, precipitating a sudden shift in the relationship's form.

My parents stayed in a marriage that was filled with communication problems and polarization. Neither was particularly happy, but it had that quality of karma where they were allowing each other to make amends to the other just by staying together. I always felt that my father needed to be loyal to my mother and allow her great freedom, and she had to allow this, forgive him, and be generous to him. Though the marriage was constantly fraught with tension, I think they did a fairly good job of it, and in their forty-ninth year together, my mother suddenly decided to leave—at age seventy-two! I had to wonder: forty-nine years? Why *now*?! But I've seen it all too often in my work—when the soul is finished, it's finished. There's no accounting for why the amount of time was spent; perhaps to the soul it's insignificant. Some invisible marker point is reached, and one person shifts out of attunement and resonance with the old vibrational pattern. At this point, the reality rapidly changes for both people. Often, the one who shifted out progresses quickly toward their destiny. The irony is that the other person simultaneously made the same decision to move on but doesn't consciously realize it and often feels like a victim.

After years of being in business with his wife, Ravi began to pursue a spiritual path. He worked with a guru back in India, read about metaphysics and the new physics, and was excited about leading-edge thinking. His wife remained traditional, with values firmly fixed in the family, material security, and Indian culture. Though he respected these things, Ravi needed to move beyond to find a new identity. "If my wife would only be interested in a little bit of it," he would moan, "then I might continue to find some common ground with her!" But she thought his spiritual focus was silly. His cultural programming held him in one role, while his home frequency was taking him in a different direction, making him into a new kind of person. At a certain point, he could bear the internal split no longer and began divorce proceedings, upsetting everyone in his family and earning himself an ugly reputation. The price he had to pay for his evolution was great, but he clearly saw that if he did not make the change, he would die. We are all evolving, and the recognition of these marker points of growth is often most evident in the way our relationships—whether personal or professional—evolve or stagnate.

When one person's wave moves out of synchronization with their partner's wave into a higher frequency, it is felt immediately. If the relationship has been relatively unconscious, the person who's "left behind" may become angry, blaming, punishing, and controlling and try to force the other back into the comfortable old pattern or even physically injure them. If the partners have made a habit of trying to stay in their home frequency, they'll have a much better chance at evolving together. They can bring the misalignment into consciousness by talking about it. "Something's different; something's changing between us. What is it? Are the terms of our original contract together—our goals and dreams—shifting? What do you want now? What do I want now? Can we occupy a new field together that radiates a higher vibration? Can we renegotiate and let our relationship adapt and embody whatever new form wants to appear from our new vibration?"

If a marriage, family, friendship, or business relationship can be fluid, openhearted, and honest, adjustments to its form can be peaceful. If the souls' purposes move out of alignment, the relationship can separate and shift to whatever form is appropriate—from a marriage to a friendship, for instance. Or the people can easily slip away, with gratitude, into totally different realities. There is really no need for devastation or anger over relationship endings anymore, since both souls are always involved in determining the relationship's form.

When You Resist Being Alone . . .

If you've been rejected or have chosen to leave a relationship, you may find yourself in a period of alone time, back at the stage of the transformation process in which you must let go, be quiet, and allow your home frequency to reemerge, bringing new ideas, motives, and connections. Lacey is a charming, sophisticated woman who was used to being in relationships with men who took care of all her needs. If she lost a relationship, she never had a problem finding another man. But as her own frequency increased, this behavior didn't work anymore. She couldn't find a suitable man. Actually, her soul was showing her an underlying pattern of emotional blockage that needed to be cleared so she could freely express her great creativity that had previously been channeled into getting men to do things for her. She railed against having to be alone and felt tortured. She obsessed about a new relationship. But through involvement with a women's group, she learned to meditate. She began attending a Unity church nearby. She gradually quieted down and allowed herself to "just be." She forgot her obsession long enough to take a painting class, found she loved it, and a new talent began to emerge. Not long after, she met an entirely different kind of man through the church who was an equal partner and championed her soul expression.

Being alone doesn't mean you're not in a relationship. First of all, you're always with yourself. And second, your soul gives you what you need—either relationships or space. When you have a relationship, the task is to relate to the soul in the other person. When you have space, you relate to your own soul and the presence of the Divine in all things. You can experience a relationship with food or television, or by going into your imagination, with your light body, your physical body, or nonphysical beings. You can love the people you don't know yet and the ones staying behind the scenes giving you some much-needed space.

The eye with which I see God is the same eye with which God sees me.

Meister Eckhart

"Relating" while being in a space phase may feel abstract at first, but by practicing mindfulness and enjoying the light, love, and generosity everywhere, you replenish and strengthen yourself; a form phase soon follows in which physical relationships reappear, renewed. So when you're alone, feel into your body, into the world, into objects, and into the air and find the home frequency everywhere. Go into harmony with it, create a love relationship with the Ideal Parent, the Unseen Lover, the Invisible Best Friend. The truth is, it's impossible to be separate or to stop resonating. You belong to a unified field of beings vibrating the same way you're vibrating. Your most valued relationships, in the end, may be with your home frequency and your next level of Self.

Relationships Have a Home Frequency, Too

Since relationships are a reinforcement and magnification of the traits and vibrations of the individual participants, it stands to reason that when two people are in their home frequencies at the same time, the relationship field will expand its capacity for love beyond what we've ever thought was normal. Remember, home frequency is openheartedness. It's feeling the way you love to feel, loving the way you love to love. When both partners are in their home frequencies, differences of personality, style, and opinion hardly matter—you *want* to like and understand each other and are always ready to laugh. It's hard to be angry at a toddler, for example, when he's in the middle of a spontaneous giggling jag—mainly because it's coming from the child's pure and joyful home frequency. It's so powerful that about all you can do is join in. Home frequencies are all that way when they're shared. They create common ground. Because everyone's home frequency is a slightly different version of love, they're all compatible, like notes in a chord.

We are all linked by a fabric of unseen connections. This fabric is constantly changing and evolving. This field is directly structured and influenced by our behavior and by our understanding.

David Bohm

Every relationship field has a home frequency that comes from the merged home frequencies, or hearts, of both people. When your relationship's field starts resonating at its home frequency, you and your partner will rarely move out of attunement and love will grow steadily. A relationship's home frequency is that state of mutual openheartedness in which you can feel the compassionate, higher sanity of why you're together and feel safe. Just think how it feels when both of you are warmhearted, authentic, and spontaneous. How does the interaction unfold? When your hearts are both open, both of you want nothing more than to support each other in becoming your best selves. When you're in a relationship's home frequency, you clearly see the frequency-matching process; you can feel the specific issues or traits that are being activated between you and understand the subtle dynamics of how the healing and clearing process will probably work and how you complement each other's destiny expression. The empathy created allows you to become more telepathic and precognitive, reading each other's minds and sensing future direction and events.

Surface issues may sometimes cause you to feel that you and your partner are incompatible (literally, not capable of compassion), but when you look for the soul vibration in the other person and value the way the relationship feels when it's in its home frequency, your heart opens and notions of incompatibility collapse. If you stay in your home frequency when the other person has forgotten theirs temporarily, it makes it much easier to shift out of negativity. As fear clears from the shared relationship field over time, the relationship increasingly becomes about the experience of soul through joy, creativity, and just plain being together.

In romantic, primary relationships, sex can play an important role in helping the partners shift out of fear-based reactions and recenter in their home frequencies. It can serve as a sacred space where the superficial concerns of the world are laid aside, and the important openhearted, trusting, cherishing, playful state of awareness can be revalidated. Most of us cannot do all the clearing, balancing, and transforming just in our imagination— we need the experience of soul saturation to be physically grounded. Sex is one of the ways to have that experience. In addition, it can balance the internal yin and yang energy for both people, and in its more refined expressions can connect both of you to a higher, unifying experience of loving and being loved. If unhealthy feeling habits are creeping into your sex life, it's especially important to clear them; you must have a reliable, safe sanctuary with another person, much as you have in your own meditation or private creative time.

The greater the home frequency in your relationship, the more quickly both of you will reveal your inner truth and access hidden talents. A home frequency relationship combines the intent of both of you to radiate your "best stuff" and becomes a supersaturated field that can materialize higher-quality experiences from a wider wisdom base for both of you. Then your relationship becomes a path to a truly awakened life.

Qualities of Home Frequency Relationships

In home frequency relationships, people:

- engage equally, make time for each other, and meet each other halfway;
- are openly affectionate, thoughtful, anticipatory, and appreciative;
- are honest and trustworthy;
- reciprocate with good deeds, adapt to each other's needs, and alternate taking leading and supporting roles as necessary;
- want the best for each other, and support the expression of each other's soul and destiny;
- dissolve unhealthy emotions and feeling habits like blame, jealousy, secrecy, and victim thinking;
- are committed to keeping their hearts open and returning to openheartedness when either or both of them backslide into fear;
- know they are both dealing with the same issues when one person brings up a problem;
- communicate about what scares, pleases, and motivates them;
- seek growth and evolve naturally through moment-to-moment sharing and experiences that stretch the participants' comfort zones;
- honor the ebb and flow of the need for space, intimacy, and social involvement.

Giving and Receiving from Your Home Frequency

When you remain centered in your home frequency, giving and receiving take on an almost thrilling, magical quality. You no longer feel obliged to give or receive; there is no more tit for tat, giving-to-get, pleasing-to-get, or controlling what and how you receive. You may find yourself giving generously without a first or second thought and receiving things you didn't even know you wanted or needed. It occurs to you to do something for your partner, and it turns out to be just what they need. They give what feels pleasurable to them, and it's your next puzzle piece. When you give and receive from a relationship's home frequency, the shared field regulates what needs to be known and done, putting ideas in your head and motives in your body.

Giving and receiving are poles of a single cycle of activity. Both giving and receiving catalyze something from the field for both people to notice and experience. Both roles—giver and receiver—facilitate self-discovery and let the partners feel themselves in an expanded way (me + the gift = a new me). I remember one Christmas when I lost a cash gift my father gave me in the pile of wrapping paper debris. After much frantic searching, I told him what had happened, but said that I was going to act as if I had the money and had used it and benefited from it, and to me it wasn't lost—so for him to please not worry. He was stunned, almost to the point of tears. Of course, an hour later, once all the worry was gone, I found the money. But we were both changed by the drama. Our appreciation for each other increased, and our understanding of the power of being both giver and receiver was clarified. Without realizing it, the way I'd received had been a gift to him.

In our transformed perception, because everything is intimately interconnected, you cannot give anything away or lose anything by giving. Giver, gift, and receiver are all aspects of you, existing in the space of you. By consciously giving or receiving, you activate *the experience the gift provides* in both of you. In truth, there is only increase, as both of you notice that both of you have the experience of the gift.

> Once individuals link together they become something different . . .
> Relationships change us, reveal us, evoke more from us. Only when we
> join with others do our gifts become visible, even to ourselves.
>
> Margaret Wheatley and Myron Kellner-Rogers

A Relationship's Home Frequency Reveals Its Purpose

When you center in a relationship's home frequency, even if it's a relationship with someone you barely know, it acts as a baseline of wisdom that you can use to find insights about why you've crossed paths or come together. From that calm, neutral, loving perspective, you can look for themes that apply to both people. After that, the shared field you both occupy will educate you by providing intuitions about how the learning and clearing process between you is likely to unfold.

Let's take another look at the folks who populate your world—even the actors and characters you notice in the media and in books. The themes they represent give clues about what the souls want you *both* to pay attention to. Remember, if one person in a pair notices something, it's relevant for both. Are your friends trying to make big money or struggling to avoid bankruptcy? You all may be ready to shift gears into a way of being successful in which fear isn't a factor. Is your partner bored and complaining about their position in life? You both may be ready for a breakthrough to a new level of self-expression. Are you reading lots of Jane Austen books? Maybe you and your partner need more romance, spunk, or a simpler life. Perhaps you're impressed by some high-quality people whose lives are balanced, productive, and making a difference in the world. You're probably ready to integrate these qualities so you can shift to an expanded phase in your own life and relationships.

What if you're in a relationship with a person who reveals a disgusting character trait? Is that terrible thing in you, too? To find the purpose in a tricky situation like this, you have to drop into underlying emotions and feeling habits. The behavior you dislike may be a way the other person "medicates" an emotional wound they're not in touch with, such as rejection or an experience of overwhelming terror. The parallel is that you may have a similar wound that you deal with in the opposite manner. For example, with two people who've been smothered by inappropriate sexual energy from a parent—whether it be actual physical sexual abuse or telepathic sexual abuse—one may become promiscuous and sexually aggressive, while the other becomes prim and proper. When they're attracted to each other, the relationship may seem crazy, and they'll likely blame the other for acting in a way that destabilizes their own position. By recentering in the relationship's home frequency, they can find their suppressed memories and common vulnerability. Then the understanding comes about how they're playing opposite roles and how they could easily be in each other's shoes. Healing can follow.

You Heal Others by Healing Yourself

Both your casual and intimate relationships contain information about how you're growing and healing, and they can teach you how to serve others and work in helping and healing professions in a more effective way. The more you include other people inside you as aspects of you, the more telepathic and empathic you'll be. The more empathic you are, the more you'll be able to pinpoint the causes of other people's pain. If you're unconsciously merged with them, it will feel like it's your pain and you'll want to get rid of it by healing and fixing *them*. The truth is that *it is your pain if you're feeling it*. Though doctors work to heal physical bodies, total healing of energetic and emotional wounds, which precipitate physical injury and illness, can only be accomplished by each person working inside the laboratory of their own life and body. "Physician, heal thyself."

Sam, who lost his parents in an auto accident when he was young, had a penchant for rescuing women who'd been abandoned. He could let himself feel his abandonment wound as it appeared in the women because it was at a safe distance, but couldn't allow himself to directly experience his own agonizing loss. His unconscious logic was that by healing the women, his wound would dissolve. Ironically, it forced the women away from him; they felt he was making them wrong by trying to change them. They then abandoned him and reinforced his wound. Healing your own wound by doing it in another person cannot work. And healing another person's wound for them cannot work. The healing won't "take" because there is no conscious choice or ownership of the experience by the other person.

Because we occupy mutually inclusive fields of awareness and energy, any healing you do in yourself heals a proportion of the wound in your partner—and everyone else in your field, for that matter—by raising the vibration that you all share. As you clear your own abandonment wound, for example, you don't help hold the other person's wound in place by matching it. Your partner can more easily choose to let go of holding their pain and making it real. Your personal vibration is the catalyst for direct body-to-body communication. *Healing is really nonpressured telepathic communication between souls and bodies*. It's as if a healthy person's body and personal vibration say to the unhealthy person, "Here's an example of a system in attunement. I'm resonating with you in love, and you can create your own version of this whenever you want." You offer your healed wound for them to experience; if they're ready, the two bodies will negotiate a transfer of the subtle pattern of how the healing could happen and how it might feel afterward. Your partner will either gratefully step into the space for healing you've provided or will leave and find someone else who matches and validates their pain. This, of course, is up to them. Being healed or being wounded is an identity choice.

> I like not only to be loved, but to be told that
> I am loved; the realm of silence is
> large enough beyond the grave.
>
> George Eliot

You Can "Read" People with Conscious Sensitivity

When you meet someone new, do you receive clear first impressions? You may notice that you pick up many subtle insights and that you know more about what makes people tick than you realize. As you first encounter someone, whether in person, by phone, or even by email, your body—your trusty barometer—expands or contracts as it adjusts to resonate to their personal vibration. You may frequency-match instantly to their emotional state. If you feel happy, empowered, or totally accepted, it probably means the other person is joyful, empowering to others, and nonjudgmental. Perhaps you feel a big slurp of sweet energy rolling over you, and realize your associate relies on charm. If you feel contracted or cold, you might immediately sense the person is tense because they're shy or worried, or they're untrusting and holding back a secret, or they need to control the details of their environment because they feel overwhelmed.

Just as people and animals can tell when someone is looking at them, your body can sense the tiny pressure changes that accompany inflated or arrested flows of subtle energy. You can feel the sensation of a dammed-up wave, a disturbed wave, a dangerously overamplified wave, or the feeling of impending polarization and opposition of energy flows. You can also feel the even, steady, consistent waves of harmonious people. Trustworthy people feel totally present, "real," and congruent. You may feel more real in their presence. When you hear truth, your personal vibration may seem to make a harmonious noise or vibrate more clearly, like a bell.

When someone lies, your personal vibration may race erratically or pause as though you've taken a short inbreath. Your body senses something out of harmony, and that can scare your body slightly at a visceral level. You might feel something similar when you encounter people who are not in tune with themselves; maybe they're saying one thing with their words and pro-jecting different internal images in their imagination, while their personal vibration seems smooth and untroubled. Something doesn't compute, and internally, you cock your head and listen more attentively. You can distin-guish many fine lines. There's a subtle difference between accelerated energy based on adrenaline (fear) and that based on healthy enthusiasm. Similarly, you can feel the difference between calm, openhearted energy and frozen stillness.

Try This!
Sense a Closed Heart vs. an Open Heart

1. How do you know when someone's heart is open or closed? Write about the subtle sensations your body and heart feel when you're with a partner who accepts, understands, cherishes, and champions you. Or when you're on a roll, playfully laughing and making jokes with some-one. How do your body and heart feel when your partner judges you, misinterprets your good intentions, blames or accuses you, becomes stoic instead of communicative, or makes decisions that involve you without asking you?

2. Recall a time when someone hurt or offended you, when one of your "buttons was pushed," and you reacted by pulling your energy in and closing down. What did your heart feel like? Recall a time when your "heart went out" to someone. Write about those subtle sensations. Practice sensing the subtle signals that tell you someone's heart is open or closed, and that your own heart is open or closed, as much as you can for a few days.

Andrew was ignored by his unavailable mother growing up and, subsequently, had a series of relationships where the women also ignored him, and he left them. He spent some years alone, working on his issues, not trusting the mechanics of how he chose women, then decided it was time, finally, to find a life partner who would really match him. One woman whose birthday was just days after his, who seemed remarkably similar to him, was particularly attractive and exciting. He could easily imagine their life together. Yet a couple relevant facts gradually surfaced: (1) she had been homeless, staying on friends' couches for nearly a year, and (2) she was still married to a man who had proceeded to marry someone else without telling the new partner of his situation. When Andrew asked why she hadn't completed the relationship, she offhandedly said, "Oh, I haven't had time to do the paperwork."

In examining why this woman would show up and why he was so drawn to her, Andrew realized that though she seemed willing to be in a relationship with him, she wasn't really available, and was exacting a weird revenge on her husband by not allowing him to be fully available to his new "wife" either. She wasn't even available to herself, since she wasn't making a home space from which she could be centered and clearly express her truth. "She must be radiating I'm-not-available pheromones, my drug of choice!" Andrew joked. He saw how he'd misread the woman's unavailability and hyperactivity as an attractive mysteriousness and that really, it was his body responding with adrenaline and dread, knowing the pattern of potential rejection only too well. So he said no to that "can of worms" and proceeded to stay in his home frequency, knowing he was very close to being able to recognize a healthy relationship.

Questions to Ask When Reading People and Relationships

- Are they congruent within themselves? Is their presentation aligned with their fear or soul?
- Are they projecting a false front or relating to me with their ego?

- Am I reading the energy accurately? Do we have any matching blind spots? Am I reacting from a hidden agenda or responding from a neutral place?
- What do they bring out in me immediately? How does my body respond to them?
- Do I pick up sensory clues: Do they "smell" fishy or sour? Or "sound" tinny, screechy, or hollow?
- What might I be learning from this person? What might I be drawing my attention to in myself?
- Are they trustworthy? Are they honest? Are they withholding or lying about something?
- Do they have a personal issue that will sabotage the relationship or the flow of a project?
- Do they have a need to control, and if so, what and how do they do it?
- Is this a beneficial relationship for me? Why?
- Is it time to update or change this relationship? Is it time to end this relationship?

How You Can Move from Conflict to Energy Flow

Part of mastering relationship resonance is understanding the cycling nature of interaction and how your various inner wave motions parallel those of your partner's. You constantly oscillate—between levels of identity, giving and receiving, clarity/love and confusion/fear, space and form, and your inner yang/assertive consciousness and inner yin/receptive consciousness. Life is not meant to be just one way. The more fluid you are—the less you stop the wave—the more balanced you are and the more you can apply wave principles to increase the flow and soul in your relationships.

Relationships can operate in the old conflict-driven, oppositional way, as represented by figures 1 and 2 in the diagram on the next page, or in a transformed way, in which frequency and energy flow play a primary role in the dynamics, shown in figure 3.

Shifting from Opposition to the Figure-Eight Flow

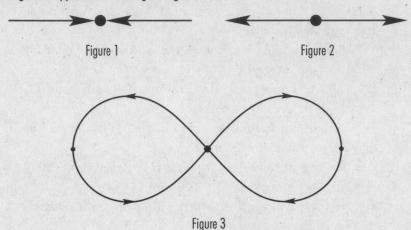

Figure 1   Figure 2

Figure 3

Figure 1 represents the trough of a wave, or the inflowing, centering phase of a cycle. Whether this applies to you as an individual or to a relationship's course of development, it is at this stage that an experience of unity, space, and gestation of the new occurs. If you're in your own private process, you'll need to pause, be still, and reconnect with your home frequency. If you're in relationship, this is when people are attracted to each other and forget about the outside world. Once the inflow occurs, the ego wants to stop the flow and have this be "the way life is." Fixations begin. In a personal process, you might feel isolated from others, cave in on yourself, feel worthless, or get stuck in depression because nothing's happening. In a relationship, the partners fixate on each other and their similarities, spend an inordinate amount of time together, virtually live in each other's lives, and lose touch with their own truth.

Figure 2 represents the crest of a wave, or the outflowing, creative phase of a cycle. Whether this applies to you as an individual or to a relationship's course of development, it is at this stage that an experience of diversity, discovery, and materialization occurs. Once the outflow begins, the ego wants to stop the flow and have *this* be "the way life is." In a personal process, when fixations start at this stage, you may feel hyperactive, overwhelmed, attached to your creations, or alienated from yourself. In a relationship, this is when partners diverge by finding out about their seeming "differences." They move away from unity into a polarized impasse where they cannot agree, fall into complaining and blaming, feel their partner isn't right for them, and end up feeling isolated.

In realities where you jerk unwillingly back and forth between being fixated in merger and fixated in dissimilarity, you create needless suffering for yourself. If you don't experience the wave's "turn," which is typically when everything you've been doing suddenly makes sense, you'll prolong the phase and this will generate negativity. By focusing on fluidity and life's natural tendency to oscillate, you can shift into a transformed way of relating to both your own higher nature and the higher nature of others. The key is seeing that (1) the flow never stops, (2) the wave or cycle always curves and continues toward a turn, (3) whatever part of the cycle you are noticing is the part you need right now, and (4) the part of the cycle you're in will evolve naturally to its next phase at the right time.

How a Relationship Can Oscillate Harmoniously

Figure 3 represents a relationship in flow, one in which you remember what it feels like to be at the end of a phase where the wave bends, carrying you into the next experience you need. Your mind doesn't freeze-frame on one position and doesn't stay stuck in either-or thinking. Instead, you trust the energy flow, notice what's new, assume it's something valuable to understand, and let yourself to be comfortable with paradox, or both-and thinking. "Being together *and* apart are both good. Having different ideas is educational." If at the most oppositional time, you can see how the trait you don't like in your partner is also in you, you "make the turn" and remember that it is possible to recenter in your relationship's home frequency.

If at the relationship's most merged times (the center point of the figure-eight), you realize that it's going to be nurturing and interesting to have your partner separate from you and see what he or she chooses to create and learn next, you won't be possessive or untrusting. If, when you're both at the most separated point in the relationship (the outer edges of the figure-eight), you realize that the thing you most want to do is "come home" and share your experience with your partner, see how they respond, and what they might add, you won't feel isolated or blaming.

... the heart in thee is the heart of all; not a valve, not a wall, not an intersection is there anywhere in nature, but one blood rolls uninterruptedly an endless circulation through all men, as the water of the globe is all one sea, and, truly seen, its tide is one.

Ralph Waldo Emerson

A relationship between two internally balanced people whose priority is staying in the relationship's home frequency can be a beautiful, harmonious dance in which both know how to lead *and* both can follow. Both enjoy being intimate *and* private; both enjoy being independent *and* social. Both trust the flow and know that what emerges from the other is meaningful to them as well. If your partner needs to be particularly dynamic, you naturally feel like holding the space for them to act effectively. Next time, your partner supports you. When your partner accuses you of something, you can be the clear one and facilitate a return to openheartedness. Next time, your partner is the calm one. Being able to continually rotate between the oppositional phases of a cycle allows you access to much more of your relationship's potential, because every completed cycle builds trust and more soul saturation of your shared field.

Formula for Successful Flow in Relationships

Phase One

1. You're simultaneously drawn to the other person because you frequency-match issues both souls want to focus on. You experience attraction, love, and admiration.

2. The love draws everything that's in the way of more love into the relationship space so it can be cleared. Both people's subconscious fears surface. You play opposite roles, have differing opinions, blame and punish each other, resuppress your problems to get along, or break up. This usually repeats, drags on, and can often end the relationship. To be successful, the relationship must progress to Phase Two and Three.

Phase Two

3. You choose to recenter in the relationship's home frequency, find a way to reopen your hearts, and communicate about what

you're feeling without blame or projection (no sentences starting with "You . . .").

4. You look for common vulnerabilities, understand your misperceptions, and choose to turn the unhealthy feeling habit *in
yourself* around (your partner chooses at their own speed).

5. You reaffirm what you value in the relationship and in the
other, restate your goals for being together, and remember that
you want to cherish the other person.

6. More interferences to greater love now surface from the subconscious minds of both people.

Phase Three

7. Repeat Phase Two! Each time love conquers fear, your bodies get
more used to the higher-frequency relationship and you stop
choosing negative vibrations because they're too boring and difficult.

Just to Recap . . .

The people in your world frequency-match with you and share a field of
vibration. It can speed your growth to understand the purposes of the
relationships in your life; other people reinforce your positive talents and
help you understand your current life lessons, introduce you to new directions, or help you see and clear emotional wounds and unhealthy feeling
habits. Fear-based relationships have particular issues to work out, but
people with highly different frequencies do not appear in each other's
lives. Relationships change or end because the partners' vibrations shift—
because the souls move into new phases. There's no need to be upset over
ended relationships if you support each other's soul growth. When you're
alone, you're still in relationship with many other things; it's impossible to
stop resonating.

Relationships have a home frequency, too, which reveals the deeper
purpose of the relationship and the mechanics of how the partners' transformation processes are likely to unfold. When you give and receive from
your home frequency in a relationship, only increase can occur for both
people. Relationships help you experience an expanded identity; you
move beyond thinking of yourself as an individual and feel yourself as a

relationship, family, group, organization, nation, and humanity. To heal others, you must clear the emotional wound you perceive in them, *in yourself*. As you become more skilled with conscious sensitivity and empathy, you pick up detailed information about others via your body and intuition. You can shift from unhealthy conflict and opposition-based relationships to healthy flow-based ones by working with wave principles to oscillate consciously.

Home Frequency Message

As I explain on page xxi in *To the Reader*, I've included these pieces of inspired writing at the end of each chapter as a way for you to shift from your normal, speedier reading mind to a deeper kind of direct experience. Through these messages, you can intentionally change your personal vibration.

The following message is meant to transport you into a way of knowing the world that's close to the way you'll experience life in the Intuition Age. To move into the *home frequency message*, just downshift to a slower, less hurried pace. Take a slow breath in, then out, and be as calm and still as possible. Let your mind be soft and receptive. Open your intuition and prepare to *feel into* the language. See if you can experience the deeper realities and feeling states that come alive *as you read*.

Your experience may take on greater dimension in direct proportion to the amount of attention you invest in the phrases. Focus on the words a few at a time, pause at the punctuation marks, and "be with" the intelligence delivering the message—live and right now—to you. You might speak the words aloud, or close your eyes and have someone else read them to you and see what effect they have on you.

REMEMBER THE SOUL'S KIND OF LOVE

Deep compatibility, soul to soul, begins with the comfortable resonance of familiarity, with the intuition that we are somehow already known to each other. As we speak, our ideas fit together like puzzle pieces, building on each other as we discover a slowly emerging cocreated world more interesting and complex than either of us could know alone. It is the excitement of our interpenetrating and interstimulating minds that fuels the romantic heart and the physical passion. The desire to discover the path into what is so deeply comfortable, the kinship that seems to preexist our knowing of each other, is the fuel for lifelong love. For the closer we come to the cause of our mysterious harmony, the more it recedes, drawing us ever outward to a larger and more expanded experience of our shared self.

The soaring and weaving of our minds stirs in us feelings of awe—that the mutuality of such exchange could be possible, that it might well last through all time. We feel stunned, our hearts melt in relief and profound gratitude, and when we anticipate new spontaneous exchanges, our hearts leap! Then, filtering down into every cell, our attention slows and moves below the level of words. We cannot speak; we can only surrender, penetrate, merge, and swoon in the beauty. Here then is the rhythm of physical attraction, slow at first, then hungrily seeking the unity of sensation, the sensation of unity. I can fall into you and let go of this thing in me that keeps me separate, that makes me a me. We are free and glad for each other's freedom.

And day by day, with faith in the surprising substance of oneself and one's partner, intimacy grows. With the rhythm of easy breathing, we loop away from each other and differ interestingly, then reconvene and exchange knowings. We grow. And by repeatedly freeing each other and desiring each other, we build a trust that reinforces all aspects of love. Everything that comes from the other is a relief that takes us Home. We make vows that are statements of higher truths, not binding contracts of self-sacrifice. We commit to believing in our partner's magnificence as a soul first, a personality second. To see the soul, the same-self awareness, looking out from the other's eyes, speaking with the other's voice—this is to collapse the false attraction of need, the false fear of blame and rejection. Seeing Soul as our daily practice makes attraction and rejection into prehistoric phenomena; we know we always were, are, and will be in each other as one another.

We vow to create and cherish the calm, warm pool of the heart-home, to recenter in this frequency and live there as constantly as we can. It becomes painful to leave this home, painful to Our Self to think negatively of the other, even more to speak negatively, even more to act from fear. The other hears our thoughts, thinks our thoughts, feels our contracted emotions, and doesn't distinguish ownership. We choose happiness and joy to serve Our Self. We commit to physical and emotional health to serve Our Self. We say Yes! to creativity and celebration as often as possible to serve Our Self. Everything we do is to "make love" and to feel love alive in us. There is no point in reminding the other of their faults; it only keeps Our Self stuck in thoughts of what isn't working. Seeing what isn't wastes precious time for bringing love through bodies.

Who are we now? Our repertoire has grown to include each other's talents and leading edges. We know in our bones what it's like to come from each other's lineages, to live each other's childhoods, to bear each other's burdens, even to die each other's deaths. Two of us living all these common elements makes it all easier to bear, easier to digest somehow. We can move faster if we want. Or, slower! New perspectives come as we evolve. Are we now more like the angels? As we relax into Our Self, we naturally reach out to include more life, invite more people into a family love with us, and know with an even bigger heart-home. We don't own each other or grasp the other; trust and appreciation pave the wide white road we travel.

8

Finding Upscale Solutions, Choices, and Plans

In order to change an existing paradigm you do not struggle
to try and change the problematic model. You create a new model
and make the old one obsolete.

R. Buckminster Fuller

Patrick is a corporate trainer and consultant who is fascinated with perception. He remembers being five years old and looking at a book while also being aware of a gaping hole on the other side of the room. He called it the "eye of God," and he simultaneously experienced being in his body *and* being inside the eye of God looking at himself with his book. It wasn't coming from his imagination, he said, but felt physical, real. Today, when he works with companies, Patrick has an uncanny ability to feel many positions and possibilities at once, to see a picture of how everything fits together and how different combinations of variables will work— almost instantly.

He says, "It's as though a huge paintbrush comes out of my center and makes a broad arc, painting a whole panorama. It's sort of like a 'video-lasso-energy stream,' a movie of energy! First it comes as a knowing, then translates into pictures at a 'mid-conscious' level. I feel the flow of the pattern and how the energy would shift if different things happened, what kinds of intentions must be held by each person for it to work, and even when someone is a poor fit. If one variable changes, the entire canvas adjusts itself. Sometimes I can slide into the way one of the participants is feeling and quickly understand how he or she thinks the process will work, then compare that to what my body knows—and I know how to communicate the nuances of the process."

I love Patrick's story because he exemplifies a new way of working with energy and consciousness when solving problems and planning. He's in the flow, open to serendipitous alignments and spontaneous choices, and allows the process to have its own intelligence. He isn't afraid of empathy, works easily with the unseen realms, reads the energy, keeps it open, doesn't try to pin down an answer too soon, and lets his intuition and sensitivity, in addition to his logic, inform him.

A New Kind of Vibrational Problem Solving Can Make Your Life Easier

Seeing yourself as vibrational and working with frequency principles changes the way you arrive at solutions, plans, and goals. In this new way of finding answers, you make decisions based on the soul's criteria, not the ego's, and don't depend on yourself alone to make every decision. You no longer define situations merely as good/advantageous or bad/problematic. In the old way of solving problems, we isolated a problem, stopped the flow, and focused all our attention on making it change into something we liked better. We gathered data and opinions, analyzed it all, projected the data into the future, chose and implemented our answer, and then ticked the problem off our list. Done! On to the next one. The old way of thinking about setting goals and making plans involved getting a vision—sometimes inspired, sometimes a smart extension of another idea—and analyzing the steps that needed to be taken to make it happen. Then we stuck to that plan through thick and thin, using mental prowess and will to power through the problems along the way. We were problem-solving and planning warriors!

Once your awareness has opened in the transformation process, you see everyone and everything as inside you, in the moment, connected, and cooperative. Answers are right there, resources appear as you need them, and you see how life is always moving you toward something better. Problems transform from things that stop your attention to things that shift your attention to a new view. In the Intuition Age, solving problems means consulting with the energy flow and the collective consciousness to see what wants to happen next. Making decisions means choosing what resonates with your home frequency and the frequencies of your relationships and group involvements. Goals, plans, and strategies aren't carved in stone but are part of an organic, constantly evolving holographic movie with which you must stay in tune. To plan, you attune to higher-frequency energy flows and merge with more complex fields, then notice what wants to happen in the expanded space. The criteria for finding *upscale solutions*—those that resonate at higher frequencies—become much more about whether a result frees the flow of soul, works in harmony with frequency principles, and produces experiences that are not just "win-win" for two sides, but "win-win-win"—benefiting everyone involved.

> It probably is only the intuitive leap that will let us
> solve problems in this complex world. This is the
> major advantage of man over computer.
>
> Tom Peters and Robert Waterman

Finding Your Destiny Solves the Soul's Only "Problem"

Before we explore the frequency principles behind problem solving and planning in general, let's look at the soul's only "problem" and the strategy for its solution. The soul's problem is this: *How do I help you shift out of separation-perception into the experience of unity?* It is the experience of separation that causes problems, after all, so when the soul's problem is solved, it eliminates all other problems. The soul's solution to separation is to demonstrate unity-perception by revealing your *destiny* and helping you live it fully. Your destiny is your highest-frequency life, where you discover your unlimited talents and get to do what you most enjoy. With destiny, you experience harmonious energy flow, perfect timing, and how the Oneness provides for you and guides each step you take. Problems become messages from your soul and problem solving becomes simply "living." You realize that life is working perfectly to bring what you need, that trying to control the future is a waste of time. The more you become your soul, the more your life becomes your destiny.

Since you're an individual in time and space, your body is a specific kind of lens, and you're naturally built to filter energy into the world in a certain way. Perhaps your filter shapes energy into architecture, or fine jewelry, or learning experiences for children, or complex organizations. Maybe you're built to travel internationally as an emissary for a cause, or to love the land and grow beautiful food. When you do what you're built for, you experience destiny, and that precipitates an experience of unity. Your destiny, as you come fully into it, makes you feel that you've been given your most cherished dream: You mean I get to do *this?* Problems? What problems? I get to do *this!* You soon see that the hidden purpose of all problems is to correct misperceptions, clear your view, and lighten you up so you can live your destiny. So if you help your soul solve its only problem, it will solve all the rest of yours.

> A problem is your chance to do your best.
>
> Duke Ellington

You Can Transform the Way You See Problems

Destiny is something you walk into gradually by facing a string of mundane problems and transforming them one by one from blockages to opportunities and guidance. As you address each problem from now on, be looking for the soul's motives. Your soul creates situations so you can learn from direct experience; it doesn't care about having money, for example, but about the experiences you create and the lessons you learn by having or not having money. When you notice a problem and look inside yourself, you'll see that you've stopped your own energy flow. Why? Your soul is indicating something you haven't fully experienced or a place where you have a misperception about the way life works. Most often, a problem is pointing your attention to an experience you need to have. So instead of looking for a strategy or action that will change the problematic situation, try relaxing into the situation to see what your soul wants you to notice.

Nancy's high level of anxiety about asking her boss for more responsibility and money, for example, is really about her desire to leave the company and become an independent consultant. Harry's irritation with the neighbors in his apartment complex, whose cigarette smoke comes into his bedroom, really pertains to his need to feel entitled to have his own home. Joan's lingering bronchial infection is a way she can subliminally access her deep grief over the death of her husband. By using their problems as stepping stones to new insights, these people can clear interferences to their soul expression and destiny.

When your mind stops on a problem, pay attention to the following things as you look for a solution:

1. **Reframe the problem so you see it as helpful.** By doing this, you take away the pressure caused by resistance and allow space for solutions to appear. Also, see the problem first as a function of your soul, not just of the physical, emotional, and mental worlds. Don't forget to center in your home frequency as you begin contemplating the problem's inner dynamics. This helps you find understanding much more quickly.

2. **Your problem is really a question. The question helps you discover an experience your soul wants you to have.** Instead of making your problem into a static statement, which locks in a reality you don't like—"I need a new car but can't afford it"—turn it into one or more questions that elicit insights about the experience your soul is trying to facilitate for you. Questions create movement. What experience will having a new car give me? What inner lesson will I learn by experiencing the renewed feelings of safety, freedom, and power that a new car will provide? What can I experience *right now* to be more confident about moving around more in the world and going to new places?

3. **Solve the problem internally first, in your imagination.** You don't have to wait to get an actual car. By imagining yourself driving a new car, feeling the power of acceleration, or the reliability of a new engine, or heightened self-esteem and pride, you're having the experience your soul wants you to have. Especially if you use all your senses while imagining, your body will recognize the experience as real. Now you've empowered yourself to come into the soul's reality where positive character traits are normal, where you truly *are* free and entitled to feel good

and abundant. The experience of driving a new car in your imagination helps you return to the feeling of what it is like to live your destiny.

4. Once you have the internal experience the problem was pointing to, the external physical problem may resolve quickly. In the case of getting a new car, once you feel the confident energy and desire to explore what a new car would facilitate for you, you might land a high-paying freelance project that gives you the money you need to buy one, or a friend may decide to get a sports car and sell her sedan, and she'll give it to you for a great price. Then you can ground the imaginary experience in physical reality and have completed a life lesson.

Sometimes problems are signs of a wrong direction, and as soon as you have the experience your soul wants you to have, the "symptom" disappears. Maybe your body starts aching and you think something is terribly wrong. The doctor can't find anything. If you relax into the problem, you find that your body just wants to move more. You start walking after dinner and the ache goes away. The soul needed your body to be capable of transmitting more energy so you could reach your next level of growth, and this was the way it got your attention.

5. Problems always indicate a "turn" of the wave and a new vibration that pertains to your destiny. Make sure you're seeing what the soul is trying to reveal. Once you give yourself the experience toward which the problem is pointing you, what new direction are you facing? What wants to happen next? What's the soul's larger intent? How has solving this problem helped open you to new, expanded possibilities?

6. If many problems occur at once, there's usually just one underlying theme. There are times when snags pile up and problems seem to breed more problems. You might feel jinxed. It's probably because there's a very big issue surfacing from your subconscious that you've resisted dealing with, and your soul is clearly telling you that it's time to take care of it. Every escape route is blocked. Even tiny things are going wrong. It's a sign that the surfacing issue is highly out of tune with your home frequency. Best not to try to solve everything, but stop and feel into the deeper tensions and questions. What experience are you afraid you'll always have to live with? What experience do you desperately want to have and know to be normal? Freeing up the core issue will often create a condition in which many problems are solved with one action.

7. Even very mundane problems contain your soul's guidance. I have some glop on my printer's roller that's making white spots on the print-outs. The plants are too dry. My computer just started turning itself off as I'm writing, and I discover the power source needs replacing. Each one of these things is symbolic of something in me that my soul wants me to notice. Is there an old way of perceiving that is blocking my ability to write more brilliantly? Do I need more emotional nurturing? Do I need to stop, rest, and rejuvenate so I can continue to do a good job? My soul wants me to tend to these "problems" in a nonproblematic way; I am to pay loving attention to each thing, to return it to its ideal state. As I do, I remind myself of my own ideal state and reaffirm that I am living according to my destiny.

> Yes, we see there are problems in the world. But we believe in a universal force that, when activated by the human heart, has the power to make all things right. Such is the divine authority of love: to renew the heart, renew the nations, and ultimately, renew the world.
>
> Marianne Williamson

What Solutions Feel Like in the Intuition Age

In the Intuition Age, solutions become revelations of a fuller way of being. They may reveal a path of action so perfect that your mind could never think of it, where latent or secondary factors and abilities become the main thing, everything you've done has a purpose, and it all weaves together with each aspect playing its happy role. Carmen grew up as the responsible oldest child protecting her sisters from a violent father. She learned to be strong and successful and manifested enough wealth to reward herself with a great lifestyle. She laughingly calls herself a "hedonist." Yet in the various branches of her work, she still focuses on healing and rescuing. Her problem is that she feels stuck and too serious and wants a different kind of life. Her logical mind came up with the answer: "I'm supposed to empower people to be strong and express themselves fully." That was a logical extension of her process to date, but it left out many of her soul's qualities that had never been celebrated as a child.

I sensed that the real direction of her new life would relate to her ability to enjoy beauty and high quality, and because she could materialize whatever she needed so effortlessly, she could infectiously communicate a pattern of how easy this could be to others. That just by being around people, and loving and appreciating them, she could help them remember what they were here to do and how they could do it. There would be no effort required and no need to rescue or protect others. By removing the feeling of "having to," Carmen could see that reveling in the pleasures of the material world was a spiritual thing and that she loved the world, she loved people, and she wanted to play with everyone. She decided to become a new kind of money coach and show people how much fun making money is. Solving Carmen's problem did not come from extending her "heavy" childhood role but from picking up a lost thread based on her joy, which she had been minimizing by calling it hedonism.

With Intuition Age solutions, the best, most natural answer simply arises into the space, perfectly meeting the need focused by the soul—and it's usually surprising in a good way. Problems are simply incoming guid-

A PROBLEM PERSISTS WHEN:	A PROBLEM TRANSFORMS WHEN:
• You judge the situation as bad and stop the flow. • You make it into a statement of truth and lock in a situation you resist. • You avoid the underlying experience the problem is aiming you toward. • You jump too far ahead, looking for the final answer. • You have an agenda that's out of alignment with your soul's intent.	• You look for the soul's reasons: what are you learning and trying to experience? • You see it as a natural turning point, choice point, or indication of forthcoming guidance and revelation. • You turn it into a question or series of questions to elicit deeper insight. • You keep the flow going by having the indicated experiences first and having faith that the realignment with soul will soon present a beneficial solution.

ance from your soul and navigational devices that indicate the need for directional shifts. Solutions often use surprising parts of your personality and odd combinations of variables, occur in miraculous ways when you least expect them, solve several problems at once, and leave you smiling in amazement.

Try This!

Relax into the Solution

Think of a problem that's affecting you. As you call it to mind, see if you can feel how it arises from somewhere deep in the center of yourself and projects out in front of you. The unconscious assumption is that the solution lies "out there" somewhere and you must hunt for it, or it must come to you. Instead, relax and let the projected problem be reeled back into the spot where it originated; let your awareness relax back into the place inside where you first became aware that there was such a thing as a "problem." Sit in there a little while; the solution exists in the very same spot. Let yourself become aware of it. It will come to mind and probably seem simple and obvious. Problem and solution exist together as an experience your soul wants you to have.

How Do You Decide?

Decisions are often seen this way: do A, B, C, or nothing. As you ponder your options, decision making may go like this: If I do A, this bad thing will happen and this good thing will happen. If I do B, that bad thing will happen and that good thing will happen. Which option has the most good and least bad? You don't want to make a snap decision based on personal biases, so you gather data and analyze it, weigh the pros and cons of each option, and project the options down a timeline into an intuited future. Kept in the realm of the mind alone, this can become confusing and paralyzing. "Where shall we retire?" One place has water shortages but lots of artists. One has nature but not much culture. One has great university towns but high altitude that would adversely affect your health. Help! In politics and organizations, of course, decisions are much more complex, yet the problems are still just constellations of smaller problems that involve many people's collective soul needs. By refusing to feel pressured in spite of a problem's supposed difficulty and centering first in the calm inner world of your home frequency—where life is simple—you can use your sensitivity to find insights that will put even complex situations in the right perspective.

How wonderful that we have met with a paradox.
Now we have some hope of making progress.

Niels Bohr

Sophia's company, powered by government contracts, began to lose business because of cutbacks. For the first time, they wouldn't make their year-end budget. What to do? Limited layoffs, no pay raises, and increased workloads for the fewer remaining employees? Pay cuts for everyone? Cut a whole program and the employees that work it? Or could she quickly summon the wherewithal to develop new income streams when she was already working a grueling schedule with no time for her family? Because other people's lives were impacted, Sophia felt she didn't have complete freedom of choice. She had to be responsible. She was caught in a stranglehold: she had a ten-year pattern of success to uphold, and yet without realizing it fully, Sophia's *inner blueprint*—the pattern of her thoughts, emotional needs, dreams, and soul motives—was shifting. Her soul was aiming her toward a new kind of work and a new pattern for her company. Her problem and gut-wrenching decisions were there to help her know it was time to reassess how she was expressing herself, but she couldn't quite see that her soul was generating the situation because the physical drama was so great and solving urgent daily problems was taking most of her attention.

I encouraged her to dip into her home frequency, suspend any fears that were coming from not wanting to fail or let people down, and check in with who she really is. She spontaneously came up with the idea of listing all the things she liked that had originally gotten her into her job. Her excitement and goal ten years ago had been about creating the biggest company of its kind in her state. Now that she'd accomplished that goal, she saw that she actually wanted the freedom to do more creative thinking, design educational programs, and advise other professionals rather than ride herd on employees and deal with government audits.

Take Time to Adjust Your Inner Blueprint

Some interesting insights surfaced as we talked. First, by carving out some breathing room to see what the process was about for her personally—instead of being a fire truck rushing to put out the fire—Sophia could consider the idea that when you change your own vibration, everything in your world changes in a parallel way. She could see the hidden purpose behind why the business didn't want to grow further in a certain direction. She could understand that if she shifted from fear to feeling what her soul wanted, great solutions—magical solutions—had a better chance of appearing to her. As she considered what excited her, she had a tantalizing inkling of a new direction that would infuse new life into her company. She sensed that opportunities would drop in her lap, as they always had. And she intuited that their big contract really would be dead in a year because the services it provided were being outdated by new trends. Her company would experience a dip, but it would correct quickly. She could feel the timeline without knowing the exact events that would occur.

> Although we believe that the logical portion of our brain guides us
> in making future decisions, studies show that humans, like animals,
> use emotions and intuition to make decisions about the future.
>
> Lynne McTaggart

The small amount of time Sophia took to consider how her inner blueprint was changing reestablished a healthy energy flow in her life. Now inspiration was moving, and she was looking forward to new developments. She didn't need to know what the new business would look like at that very moment, but feeling relaxed about it gave her more space and freedom to engage her imagination. She was able to meet with her comptroller and devise a plan for gradually preparing for the end of the big contract in a way in which the employees servicing it could be trained to work in other areas, knowing there would be a period of belt-tightening. She knew the ultimate solution would come via a series of decisions, not one huge immediate one. Each decision would carry the core vibration of what was most-real to her, what was most attuned to her home frequency. That way, her organization and employees would evolve to stay in resonance with her. She'd be re-creating her business from a more accurate vision.

Try This!

Update Your Inner Blueprint

Your life may be comfortable or uncomfortable; either way, you may not be consciously up to date with the latest version of your inner blueprint, since it constantly evolves. To sense what your soul wants you to do next, write about the first impressions you receive in response to the following questions:

- Is there a dream from earlier in your life that you've let slide? Is there a more current version of it now that might still be intriguing to you?
- Is there a life dream you've always thought you would achieve, but in truth, you're no longer wedded to it? If you let it go and open fresh space, what do you see yourself doing instead?
- Where are you bored in your life? What ideas and curiosities are tickling the edges of your consciousness, beckoning you to become more involved?
- Is a course of action or a goal actually complete? If you let it go and open fresh space, what do you see yourself doing next?
- If age or financial resources were not limiting factors, and someone else helped facilitate a new reality for you, what alternate existence might you like to explore?

These questions can open your mind enough to allow new opportunities to occur. The most important part of aligning with your most current inner blueprint is to welcome change and new ideas, then make authentic choices in each moment about what fascinates you, what produces enthusiasm and engagement, and what makes you feel proud of yourself.

Weigh Your Options Using Conscious Sensitivity

As you remove barriers to a higher personal vibration, you become more aware of the wide variety of vibrations in the world. You also see how many options you have—as my friend's three-year-old said the other day when she told him to do something: "But I have *choices!*" Now that you see it all, how do you decide what to do? Sophia's way of progressing was just right; she made a commitment to check with her deep truth on every decision she made. It's important to remember the wonderful power of the present moment to act as a filter: you only need to consider one idea in each moment, and in that space, you can feel into it fully. In fact, you can ask your Inner Perceiver to bring you the options and ideas you need, in the order you need them. That way you can easily pick your way, step by step, through the maze of confusing data.

Part of weighing your options is using your truth and anxiety signals and your conscious sensitivity to notice layers of insight and meaning in what presents itself to you. For example, someone might advise you to take a specific plan of action, but your subtle sensitivity tells you the answer is not complete enough. Do you need more information? A deeper perspective? Is the solution missing a key step? Is this one piece of a larger solution that will fit with a couple of other people's perspectives? You actually know much more at a visceral level than you consciously realize. If you practice making this subtle information conscious, you'll shortcut your decision-making process.

Try This!
How Does Conscious Sensitivity Help You Decide?

Pay attention to the way you notice subtle layers of meaning. Feel into each of the following from a quiet, centered place, and see how your sensitivity helps you get to the tipping point where you can decide with a corresponding feeling of "deep comfort." For example, you might experience a knot in the pit of your stomach when it's not right for you to take a new job offer. You might find your body seems to act independently of your mind sometimes, when something urgently needs to be said or done. How does your sensitivity help you decide:

- Whether a solution is "elegant" and will solve several problems simultaneously?
- Whether an answer doesn't go deep enough or is missing important information?

- Whether a path of action is wrong or has a high possibility of failure or danger?
- Whether a situation is being forced or timing is off?
- When to take action? When to wait? What to do first? Second? Third?
- When to make an issue of something or let it roll off your back?
- When to complete something and when to drop it and move on?

The more you work with your conscious sensitivity—and that means pinpointing the very refined signals your body gives you—the easier it will be to find upscale solutions and make good decisions.

You Can Plan and Set Goals Using Natural Wave Motion

The energy in your body is accelerating so much that change and evolution are now becoming more stable than physical form. Ideas don't lock in well anymore, and a solution, once determined, may only last a short while—sometimes days instead of months and years. Think of it like this: every second, you're making a choice. Stand up now, eat now, go left, go right. Meanwhile, everyone else is choosing, too. Your choices affect other people; theirs affect you. It's a huge constellation of choices evolving simultaneously like some amazingly complex holographic video game that everyone's playing at once. It's possible that a year ago my life plan included traveling extensively overseas giving seminars, but now someone else has been moved into that position, and the world doesn't need me to do it. Instead, "we" need me to write more books. If I'd used the first vision as a big part of my business plan and never checked for an update, I'd be mighty disappointed and maybe even think something was wrong with me when the overseas jobs didn't materialize.

> When you change the way you look at things,
> the things you look at change.
>
> Max Planck

The trick with planning today is to stay focused in the present moment. You never need to leave the moment, because it's expandable. Think of it as an clastic ball that breathes with you. As you "inhale," or stretch your mind, the moment expands to include more—morc of the timeline, more space, more collective wisdom, more energy. It's during a phase like this that you access the vision, or *inner blueprint*, for your destiny and see how everything is coordinated and how we all help each other. When you "exhale," or concentrate your mind on a single focus, the moment contracts and includes less time, space, and visionary overview. Now you feel your body and personality and are motivated to take a physical action toward a specific goal. Each time you expand the present moment, it's like opening a wide-angle lens; when you contract back to your personal reality, it's like zooming in on a tighter, more detailed view. You constantly breathe, moving in and out, including more or less of the universal body of knowledge in your present moment.

> The human soul has still greater need of the ideal
> than of the real. It is by the real that we exist;
> it is by the ideal that we live.
>
> Victor Hugo

It's hard to hold both the individual reality and collective consciousness reality in mind at the same time, so it helps to rock intentionally between them. You expand out and check your vision, receiving an impression of what your potential reality can be. Then you contract back to your body and receive a specific goal or an urge to act in a particular way. After you take the action and complete the goal, you reenter the blank space at the end of the wave and expand back out to the vision again. Each time you check your vision, it's new. *Your* vision is part of the Plan for *everyone*, and it's evolved since you last visited it a few years, days, or hours ago. By checking in often, I might notice that my desires have mysteriously shifted; I really don't want to travel overseas anymore—it would feel better to stay home, be cozy, and write a new book. The field of everyone's needs and decisions is helping me do what we all need me to do, and it comes to me as my best, happiest decision. I don't feel it's a sacrifice to not travel overseas; I'm happy for someone else to do it. When you receive your updated inner blueprint, it always feels like a natural, interesting next step.

It used to work to find your vision and hold it for, say, five years as a business plan, and not question its relevance. Sophia is a good example. She created a workable plan for her company's growth, and it was purposeful and successful for ten years. But now, the way she structured her business growth has been eclipsed by a new way of thinking and acting. Her challenges are: (1) realizing this, (2) realigning herself with an updated vision, and (3) redesigning her methods and systems to reflect it. She originally expanded to receive her vision, then focused and materialized it but didn't expand back out to check the vision on a regular basis because it was working just fine. Therefore, she was surprised when the structure began to be outmoded. If you check your vision often, you'll stay abreast of change and won't be disoriented or derailed by it.

Your Experience of Time and Timelines Is Changing

As you move into the fluidity of the Intuition Age, you'll notice that your perception of time changes. At first, you feel the world going lickety-split and that you must squeeze more into fewer hours and rush to keep up. You finish one thing and there are ten more calling for attention; you never have time to enjoy your successes and appreciate your own creativity or that of others. The items on your "to do" list are crowded into your near future. You can't help thinking back to how quiet, beautiful, well-paced, and gracious it was in the "good old days." Your mind is jumping into the past to rest and into the future to orchestrate plans. You're barely in the present at all. Without being in the moment, you feel pressured, can't feel your home frequency, and don't experience your limitless energy reserves; it's easy to become exhausted and drained. In the old perception, your experience of time is a function of relativity—comparing past to now and future to now.

What's happening in the Intuition Age is just the opposite—past and future are being swallowed up by an expanding present moment, so there is no more relativity and comparison, just greater presence and saturation of awareness into *one* eternal now-moment that contains everything you need. And when everything is in the now, and happens *Now!*, energy moves to light speed. The lag time it used to take to cross the gap you had imagined between present and future, between action and result, no longer exists. So when you're solving a problem, making a decision, or setting goals, you need to factor this in—results can occur much faster, without as much of a linear, logical process as was once required, especially if the people involved all understand this. Oddly enough, once you pass through the experience of the speedy "jump to hyperspace" and enter the expanded Now, you settle into your home frequency and life becomes calm and almost timeless. It's not fast *or* slow; it modulates in response to your energy level and emotional experience. All the tasks miraculously get done. Materializing becomes supremely efficient because life cooperates with itself, people help you, and synchronicity and coincidence become normal. The more you hold the old habit of keeping the future and past separate from yourself, the more stress you create for yourself and the more "problems" you experience.

By rocking consciously between your personal reality and higher vision, you'll develop a natural sense about timelines and trends. You'll be more inclusive, alert to potential setbacks before they occur, and aware of inter-connections between more variables. You'll sense the probability for one path to take priority over others. You'll start with a snapshot of what's most probable that you'll hold loosely, and you'll gently adapt your plans to the shifting energy flow, allowing the final outcome to be what it needs to be. You'll enjoy the process of skillfully navigating the curves and dips in the process, the accelerations and decelerations, the places where a new strand weaves in or one is dropped from the mix. The timeline is a living thing, reflecting the needs of everyone involved in an organically chang-ing way.

Try This!
Feel the Path of a Timeline

1. Think of a project or process you're about to undertake. It might be a trip, or the development of a new product, or how you're going to decide what college to attend. Go into your centered, quiet, home fre-quency space and bring the entire process—from first glimmer to final reality—into the space with you. It all exists in your expanded self; part of you has already experienced it. Relax.

2. Feel into the energy of the process and let your body's truth and anxiety signals show you where things flow or fixate. You are sliding along the process, like you're floating downstream, experiencing every point along the way. Receive impressions about how the flow modulates as it progresses. Here's where the current slows; here's where it speeds up. Here's where something gets snagged, and here's where new people and experiences weave in. Here's where the flow dries up for a while. Here's an event—here's an even bigger event.

3. Draw the timeline, like a river, on paper. Let it widen, narrow, bunch up, and smooth out. Indicate different sorts of wave action or energy at points along its course. Indicate events as large or small dots. Feel into what might be the cause of these flow changes and put labels on the drawing: like "argument," "loss," "positive energy," "synchronicity/good luck," or "stall/setback."

4. Update, reexperience, and redraw the timeline at regular intervals to stay in tune with how it wants to evolve.

You Can Notice Problems Before They Occur

Because your mind likes definition and stability, it's easy to ignore the early signs of a shifting direction in your energy flow, which in the old way of thinking we used to call "trouble brewing." By using conscious sensitivity, you can be more alert to the threshold of visibility for a change and avert problems before they occur. If you pay attention to the way you're feeling, the way your employees are behaving, or the signs you're seeing in your environment, you can notice the effects of various wave actions that accurately predict coming situations. You might sense life beginning to sputter. You might notice clogging and snags.

> Coming events cast their shadow before them.
>
> Ancient Proverb

Let's go back to Sophia's situation: She realized in hindsight that she had some early warning signals. She was working longer hours to keep up with what everyone else needed from her. She'd been feeling ten times more stressed than ever before and was often on the verge of getting sick.

For the first time, she hired an office manager, who as it turned out couldn't do the job, which set the company back months correcting the mistakes that had been made. She moved her office to a larger space and encountered an inordinate number of snags getting it and the computers up and running smoothly. Whenever you feel you're pushing or pulling, or if there's a contraction or knot of energy in your body or environment, when your energy is taxed and your joy goes underground, or if your health is suffering and the same effort doesn't elicit the results it used to, change is in the works!

Try This!
How Do You Sense Change Brewing?

1. Recall a time when you encountered a period of trouble and problems. Were there early warning signals? If you'd been more sensitive, how might you have noticed that a shift was in process? What did it feel like in your body and environment just before the problem(s) surfaced? What actions might you have taken if you'd realized what was trying to happen? Feel into the following to see:
   - What senses do you use when noticing a problem? Where does it register in your body?
   - How do you know if something is going to be a serious problem?
   - How do you know if it's a problem now, or a potential problem, and how long the shifting process is likely to take?
   - How does a "just-right" solution feel?
2. You might practice, from today forward, sensing energy ebbs and flows and times when your mind glosses over a snag; how you recognize that it's time to check your vision; what action to take and when; and whether more information is necessary to arrive at a fuller perspective.

How Do You Recognize Upscale Solutions?

Upscale solutions are those that promote the highest level of soul expression for all the people affected. They are based on the interconnectedness of life and therefore do not take shortcuts to alleviate immediate tension that will cause damage in the long run. It is understood that we exist in a living system and that damage to one part handicaps and hurts every other part in some way—and this includes the earth itself and all forms of creation. There is no place in an upscale solution for negative emotion and fear-based behavior, such as meanness, revenge, secrecy, cheating, lying, or stealing.

An upscale solution always facilitates healing and truth. It highlights places where there are underlying misperceptions—places where universal principles have been misunderstood—and corrects them. An upscale solution helps important information surface and helps shift the mind from either-or, right-or-wrong, black-or-white thinking to inclusive, both-and thinking, where fresh creative flows are precipitated by considering paradox and complexity. An upscale solution often jumps up a level and eclipses the old model of the problem with a more comprehensive model, like including a solution's spiritual impact, for example. To find an upscale solution, ask yourself these questions:

1. Is this problem the leading edge of a new piece of learning or a new opportunity to grow? What is it trying to contribute to your life or your group's experience?
2. If the problem remains in place and the process continues in the direction it's headed, what's likely to happen? If you let the system adjust itself naturally and reorient, what happens?
3. If the worst possible scenario occurs, what do you learn? Do you need to do it that way, or can you learn the lesson in your inner world via imagination? Even when there is "failure," how do things reach a positive flow again?
4. How does the energy flow want to move things into a better pattern? What supersolution is trying to happen, independent of your will and good ideas?
5. What solution has the most "deep comfort," joy, and effortless materialization of highest results at the physical, emotional, mental, and spiritual levels—for individuals, teams, organization, clients, community, environment, and future generations? What is the best win-win-win snapshot in this moment?
6. Can the means (process of fixing the problem) be as valuable as the end (outcome)?

7. Are you looking at too small a piece of the process as the problem area? Can you understand a multilevel (physical, emotional, mental, and spiritual) cause?
8. When you can't find a solution, can you give the problem to your body and the unified field; can you ask for help from the unseen realms and recognize insights when they come?

Try This!

Find an Upscale Solution

1. Select a problem you're currently concerned about. List the possible solutions you've come up with.
2. Subject your problem and possible solutions to the eight considerations listed in the previous list. Make notes in your journal.
3. In addition to the eight considerations, examine your solutions for too much willpower or impulsiveness, self-sacrifice, harm to others or the earth, speed at the expense of the body's ability to experience reality and safety, and motivation driven by fear or outdated reasons. If these were eliminated, what might the solution look like?
4. Is a vital piece of information or insight missing? Ask to receive an intuitive impression of what it might be and how it might affect the solution and its timing.

When you finish asking and answering all the questions, your upscale solution will feel like a joyful relief and will bring renewed energy.

> It's not that I'm so smart, it's just that I stay with problems longer.
>
> Albert Einstein

Just to Recap . . .

In vibrational problem-solving and planning, you see problems as questions that point toward an event you need to have so you can experience more of your soul and destiny. Solving problems means seeing what wants to happen to reestablish a healthy energy flow and moving through the turn of your wave to find new direction. Problems occur when you perceive separation. Finding your destiny is the ultimate solution in life, because when you're living as your soul, you experience unity, and subsequent problems and planning for the future disappear. Solutions in the Intuition Age feel like the "perfect fit" and always free soul expression and enthusiasm. Upscale solutions are win-win-win, serving all forms of life, and address the multiple levels of human experience—physical, emotional, mental, and spiritual.

Making decisions, even in complex situations, requires a brief period of recentering in your home frequency to check your inner blueprint or vision. Your life direction may be evolving. When you have too many choices, you can call on your body's subtle truth and anxiety signals to weigh options. When planning and setting goals, it's important to oscillate intentionally and often between your personal daily reality and your highest vision. Your destiny evolves along with everyone else's, so you can't lock in plans for very long; they must be fluid and in the moment. Problem solving and planning in the Intuition Age change your experience of time and working with time-lines, as you realize that past and future are being absorbed into an expanded present moment, accelerating the materialization process. Timelines are affected by the choices and plans of all people and must be checked often.

Home Frequency Message

As I explain on page xxi in *To the Reader*, I've included these pieces of inspired writing at the end of each chapter as a way for you to shift from your normal, speedier reading mind to a deeper kind of direct experience. Through these messages, you can intentionally change your personal vibration.

The following message is meant to transport you into a way of knowing the world that's close to the way you'll experience life in the Intuition Age. To move into the *home frequency message*, just downshift to a slower, less hurried pace. Take a slow breath in, then out, and be as calm and still as possible. Let your mind be soft and receptive. Open your intuition and prepare to *feel into* the language. See if you can experience the deeper realities and feeling states that come alive *as you read*.

Your experience may take on greater dimension in direct proportion to the amount of attention you invest in the phrases. Focus on the words a few at a time, pause at the punctuation marks, and "be with" the intelligence delivering the message—live and right now—to you. You might speak the words aloud, or close your eyes and have someone else read them to you and see what effect they have on you.

FLOW SO YOUR PROBLEMS DISSOLVE

Living is moving. The roller-coaster takes you from wave to particle, from movement to pause. The wave brings flow and the joy of release. The pause makes you aware of self—your individuality, plurality, and unity. Wave and pause. Expand and recenter. Separate and reunite. Give and receive. Communicate and feel your connection. Learn and experience wisdom. You are the wave and you are the particle and you are the wave again and the particle again. Each time, new.

You have learned an unhealthy habit of using the mind incorrectly. You are in a wave, and the mind states: "Life is energy, and I am motion." Then, when you pause and become a particle, the mind thinks it must redescribe: "Life is solid, and I am a finite individual." When you move on to be a wave again, the mind feels pressured to change the definition back to "flow." It is stubborn in its habit of doing this, of wanting you and life to be just one way, of not broadening enough to experience the whole particle-wave nature of life and self. With its narrow view, each time the mind reaches the turn of wave-flow into particle-pause and back again, it says, "I am wrong. Something is wrong. I have a problem. I don't like this. I must control this change." Fear is generated, and needless disturbance ensues.

Focusing on having a problem, when instead you could be enjoying the newest phase in your experience of self, is the great folly of humankind. Under the feeling of having a problem is a subtle feeling that indicates you are shifting to your next, even more enjoyable experience of self, be it flow or pause. Enter that experience without labels and find communion with it. Move, express, and create or pause, recenter, and appreciate. Then watch the external world—your dream, your movie—adjust itself naturally to match the energy cycle. Without the mind's hesitation, the world has no hesitation. Problems and solutions dissolve; appearance and disappearance and reappearance occur instead. Life's forms come and go and evolve.

You can teach your mind to stop defining and "be with" this moment's experience instead. Show your mind how to recognize the feeling of the turns of your cycle, to attune to the pleasures of both phases, so it may embrace them happily and release the concept of "wrong." Whenever you notice your mind feeling separate or resistant, help it drop in, rejoin the current phase, and look for the pleasure. Soon, deep pleasure will be the unifying principle that lets you bring love through every precious moment of awareness. With the soul's pleasure as your organizing principle, there can be nothing wrong, no problems, no solutions. Life evolves itself wisely. Answers are momentary stopping points where mind meets soul, and in these meetings, there is the joy of choosing to create the next appropriate thing. You can give up having problems. You are without need when you relax into the Now.

9

Creating a High-Frequency Life

Through fear of knowing who we really are we sidestep our own destiny,
which leaves us hungry in a famine of our own making . . . we end up
living numb, passionless lives, disconnected from our soul's true purpose.
But when you have the courage to shape your life from the essence
of who you are, you ignite, becoming truly alive.

Dawna Markova

The increasingly popular trend toward thinking we can intentionally create or materialize our life, or things in our life, is fairly radical, really, when you consider history's long emphasis on people being helpless victims of the gods and their fate. It means we've evolved to a level of knowing where we can see how our internal states of spirit, thought, and emotion affect our physical reality—how the nonphysical actually gives rise to the physical. *Materializing*—what many people refer to as manifesting—is the idea that one can, by force of will, desire, and focused energy, make something come true on the physical level. This is knowledge that for ages has only been available to avatars, magicians, and alchemists. But not long ago, the *law of attraction* surfaced, bringing previously esoteric and metaphysical ideas into the mainstream. It showed that we are more than ready to understand how the power of personal vibration helps shape reality. The law of attraction basically states that "like attracts like," "you create what you vibrate," and "being happy helps you materialize the life you want." This, in my experience, is mainly true, but it is a first level of understanding. Let's explore more of the nuances of how materializing—and the reverse process, *dematerializing*—work.

There's More to Creating than the Law of Attraction

At a time when society's transformation process is hovering around stage 3 (emptying Pandora's box, or the subconscious mind), stage 4 (retrenching, resisting, and resuppressing subliminal fears), and stage 5 (old structures breaking down and dissolving), many of us are strenuously trying to avoid the chaos caused by loss and letting go. We want control over our lives and think money and possessions are the answer. To date, the law of attraction, which actually addresses a principle of spiritual growth, has been skewed a bit too much toward overcoming insecurity by attracting material success and relationships—and that means the principles are predominantly being applied out of fear. To progress into the more advanced stages of transformation, you'll need to penetrate further into the frequency principles that underlie the law of attraction.

One of the hallmarks of having passed through stage 6 of the transformation process—where you stop, recenter in your soul's love, and let your home frequency reprogram your personal vibration—is that you discover a new world based on unity. You experience yourself as energy and part of a unified field of vibration. You directly experience how everything exists in the present moment as a *superposition*, or potential reality and how all beings cooperate to materialize the world according to the highest, most current vision moment to moment. This new perspective means your goals need contain no desperation, and a high-frequency life or destiny, where soul expression is easy, is your birthright. As a result, your understanding of the creation process and success changes dramatically.

I don't know about you, but the more I learn to maintain the new perspective of the Intuition Age, the easier it is to hear the false notes. Some days I feel bombarded by the rather forced cries of the "marketeers" hawking internet sales strategies, coaching-for-success services, get-rich-quick schemes, and one-of-a-kind seminars that will change my life. Many of them advocate law of attraction principles without having moved very far through the transformation process themselves. If I'm not alert, it's easy to become exhausted dealing with society's need for celebrity, fast money, and material possessions. I find myself rebounding off the strained voices and gravitating to the peaceful ones who radiate the joy of creating *with* each other for the pleasure of expressing themselves harmoniously from their home frequencies with win-win-win outcomes. One way screeches, the other makes beautiful music. There's a challenge here: to see past superficial success; find our way into our destiny as a new kind of success, and to understand materializing and dematerializing as a normal part of our awareness repertoire.

World conditions challenge us to look beyond the status quo for responses to the pain of our times. We look to powers within as well as to powers without. A new, spiritually based social activism is beginning to assert itself. It stems not from hating what is wrong and trying to fight it, but from loving what could be and making the commitment to bring it forth.

Marianne Williamson

It's important today that in addition to questioning and redefining success, we also pay attention to the way we describe what happens during the creation process, that we no longer use linear models to depict processes that are holographic and instead develop models that accurately embody unified field dynamics. Linear models are based on the idea that we're separate, while holographic models are based on the unity of the moment and field that we all share. Keeping your mental models up to date will streamline your ability to materialize and dematerialize your world. Why? Because *how* you believe things occur is the inner blueprint beneath the blueprint of *what* you want to occur. If either blueprint is misaligned with higher truth, you'll experience snags and distorted results. Thinking that you need to "attract" material abundance, for example, is a slightly outdated model containing a few inherent misperceptions.

First, material abundance may or may not be your soul's real goal. Second, in the holographic way of perceiving, you and the thing you want are in the same moment and space, so you can't actually attract it because you already have it. Attraction is an idea that comes from the linear mindset of separation: what you want is "over there," and you must make it come to you with will and cleverness. In a unified world, instead of attracting something, you notice it *already existing in your reality*, and you keep paying attention to it. No effort required. Third, because you use will when attracting something, no matter how happy you make your personal vibration, you instantly disconnect from the experiences of unity, love, being provided for, and how the universe is always win-win-win. When willpower surfaces, the ego is back, thinking it's running the show, reinforcing the idea of isolation and difficulty. It's a setback for the soul because this subtle willfulness means that you've stopped trusting that the soul is bringing exactly what you need. Therefore, energy is wasted in protesting, over-riding, and trying to change what you have before you fully understand the perfection of why it's there.

> The spirit is the master, imagination the tool,
> and the body the plastic material.
>
> Paracelsus

Here's How Materialization Begins

The first level of materialization comes from your soul, independent of your conscious mind; the soul materializes the important situations you need for your life's growth plan, always in concert with other souls. You can go along with the plan or resist it, make mistakes, and stall along the way, but the soul's plan automatically self-corrects, like a stream finding new channels around a logjam. What's necessary will keep reappearing until you engage with it fully. Your life plan is loosely mapped out and adjusts itself continually as everyone evolves. Do you need a certain relationship at a certain time in your life? The souls arrange to meet. Do you need a new opportunity? Maybe you need a loss. The soul arranges the circumstances. When it's time for a breakthrough, virtually nothing—no amount of mental and emotional density—can stand in the way of a soul whose time has come! Your circumstances are the soul's means for creating experiences that help you learn your life lessons, and they can change quickly, like a mirage in a heat wave, in response to the soul's intent.

Basically, your world materializes out of who you are. It is an extension—a lower octave—of your personal vibration. This means it's a mixture of love and fear. Part of your soul's light and truth, or your high home frequency, gets through your sieve of fixed ideas and fear-based emotions and radiates out to create your world. But part of your personal vibration, and thus your materialized reality, comes from the low-frequency shadows cast by your blockages, as they, too, radiate through your field. It is your materialized shadows that you interpret as problems, and it is in clearing them that you learn your life lessons. To begin creating a high-frequency life, first stop materializing needless negative situations.

Your mind and feelings play a secondary role in materializing, but largely, they either serve to block and stall, or ease and speed, the soul's intentions; it all depends on what you choose to pay attention to and make real—your loving soul or your fearful ego. Eventually, when fear finally dissolves, your mind and feelings merge to form a sublime sensitivity and become a perfectly clear transmitting lens. Then your personality's desires and soul's intent become identical. No willpower is needed to "get" what you want because the next just-right thing is always showing up. In this frame of mind, the process of creation and dissolution is characterized by a high degree of playfulness and innocence, and because of the lightness, you're free to materialize a wide variety of experiences that all flow harmoniously with the evolving destiny of humanity. A high-frequency life, then, has a natural quality with great fluidity and pleasure in simple things.

Attention Causes Energy to Materialize

It is attention—interest—that causes energy to materialize. It begins when you notice an idea you want to experience already existing in your reality. Because you notice it and call it out from the field, it starts to appear as an image in your "image–ination" and it takes on a kind of psychic weight. The longer you pay attention and invest energy into the idea, the denser and slower the idea's vibration becomes as it moves down through the octaves of thought and feeling into form. The materialized idea doesn't journey to you across distant quadrants of time and space; it simply resonates down through your personal vibration, passing through the three levels of your brain, gradually taking on shape and sensory data, becoming more real to your body until it's perceived through the filter of your reptile brain as something real. As it materializes, the idea catalyzes out of your personal field as a new, smaller field vibrating at the frequency of matter, within arm's reach at the moment you need it. Materializing life is actually a function of communion, of feeling into and merging into an idea with all the octaves of your awareness—spiritual, mental, emotional, and physical—until you and the idea exist together in three-dimensional, "solid" reality. The materialized idea appears as a function of love, coming because you want it, and disappearing back into the unified field to give you space when you need it.

Everything that happens in all material, living, mental, or even spiritual processes involves the transformation of energy ... Every thought, every sensation, every emotion is produced by energy exchanges.

J. G. Bennett

Whatever you materialize embodies *your* personal vibration in some way. If you hold the idea of self-sacrifice, your materialized world holds self-sacrifice to the same degree. In this case, you might experience people who are victims, or you might have losses, betrayals, or relationships with self-centered people who don't appreciate you. If you're joyful, the life that appears for you will be entertaining. If you hold an idea that you have to work hard to make ends meet, you may materialize job opportunities with low pay or long hours. If you feel privileged and cared for, you might materialize an inheritance. Similarly, illnesses and injuries correspond with inner blueprints based on negative emotions that relate to the afflicted body part. For example, lung problems often relate to grief, neck problems to stubbornness and lack of trust, leg problems to hesitation to take a stand or move forward. It's not that you "attract" health problems or negative circumstances because you're "bad"—it's that somewhere in your personal vibration, there's an all-too-human unhealthy feeling habit, such as not feeling respected or being afraid of being punished or abandoned.

For something to fully materialize, your body needs to recognize it as real. That means your senses must be activated and the new object, situation, or experience must feel normal. I can touch my computer right now—it's only inches away—but I remember when I decided to upgrade to my fancy Macintosh. It began as a thought and progressed to an image. I imagined using it, felt the reality of how I would pay for it, went to the showroom and experienced a demonstration and played with one, touched it for real, and got excited. I could feel how it would improve my life. I could see it on my desk. I absolutely knew it was coming to my desk. And I set the process in motion to buy it. Now my body is convinced that this level of sophistication, this speed, this size monitor, are all normal. When it's time for another growth spurt, I'll get restless, bored with it, distracted by ads for the next new model, and go through the process again.

You Do Not Create Alone

My friend Anne and I did a meditation experiment together, sitting face to face, eyes closed. We imagined that something would appear in the space between us that we wanted to materialize. She decided on an amount of money with one more zero on the end than I would have been comfortable suggesting. We asked for it to appear, like a hologram. I asked her to describe the money aloud in detail. As she did, I visualized it. It was showing up in bundles of bills, so I asked her to open the bundles and touch it, to describe the sensations. I imagined experiencing the same thing, and if I got an extra sensation I described it and she imagined it, too. We went on this way, smelling the money, spreading it and restacking it, both of us putting light on the money in our minds. Then we thanked it and let it dissolve.

I thought nothing of it for weeks, as it was beyond my normal level of short-term income and not likely to appear. Yet in an unexpected occurrence, on a trip to visit my publisher to talk about the proposal for this book, I was surprised by their announcement that they had already accepted the book for publication and were able to offer me an advance. It was the exact amount from the meditation! Anne's process took a little longer, but within a few months, her clientele mysteriously increased in spite of a falling economy, and she made the money. The success of the exercise, for both of us, acted as a sign from our souls that we were moving into an expanded phase of self-expression. We had had no fear motivating us to do the meditation and did not rush back to do it again. We did observe that the effect of both of us simultaneously bearing witness to the reality of the money in a tangible way and having our bodies reinforce the reality to each other through a common resonance may have helped the outcome occur.

You do not create alone. Everything you materialize is cocreated with other souls and even with the cooperation of waves and particles. And we are all cocreators with the mysterious, beneficent Divine. Sometimes I like to visualize a group of particles responding to my need, saying, "Yes! Let's hover together in Penney's reality for a while and shape ourselves into a new computer for her!" The truth is, we are all so agreeable that everything you want affects what everyone else wants. As you become clear, for example, about wanting to take a trip to Greece, the other souls who can help this happen will shift their desires naturally, so they'll be ready and available to share the experience with you and help you materialize it. It's for their benefit, too.

There's plenty of room to materialize what you want. By having a desire, you don't deprive someone else; our collective desires shift, that's all, so we all want what we all want. Your desires may well be seeded into your awareness by the needs of others, so with this in mind, it makes sense that lack of clarity and wishy-washy decision making might not be the most helpful thing to contribute to your fellow souls. And it may stall everyone's growth to some degree when one of us holds back, refuses to grow, thinks we're doing it all alone, or hoards resources as though there isn't enough. In the Intuition Age, there are subtle ways of using your awareness compassionately that serve or hinder the soul's work of materializing a clarity- and love-based world.

> We are the miracle of force and matter making itself over into
> imagination and will. Incredible. The Life Force experimenting with forms.
> You for one. Me for another. The Universe has shouted itself alive.
> We are one of the shouts.
>
> Ray Bradbury

How to Work Intentionally with Materializing and Dematerializing

When you play with creating a high-frequency life and materializing and dematerializing forms and situations, there are some steps that help the process flow more easily.

1. **Take stock of what you have; appreciate and use it; this is what you already created.** Really good new things won't come until you use what you've been given. If you eat a great meal, you don't pile another great meal on top until the first is digested. What you've materialized is like a solution to a previous problem; you must find the experience that the forms exist to provide for you and integrate it. If you want to move, have you loved and appreciated your current home enough? What kind of experience does it exist to provide for you? What are you giving back to it?

2. **Clear unhealthy feeling habits and negative thinking, and dissolve any existing forms that are in the way of your new expansion.** Do you want a new house? You may need to eliminate blockages about what you think you deserve and let go of the old house first. Dematerializing a reality can be just as much fun as materializing one. When you need something to disappear, pull your invested attention out of it. Boredom is your friend. When a relationship needs to end before you can move on, forgiveness and appreciation can help complete the experience. If you need to leave a job, lack of motivation can work in your favor. Dissolving a reality means digesting the value you've received, shifting your attention into forgetfulness, then just being for a while. Pulling invested energy out of an idea lightens it and it becomes more transparent and ghostlike, eventually disappearing.

3. **Collect your energy, recenter in your soul's love, attune to your home frequency, and check to see what your highest vision looks and feels like.** By feeling the way you love to feel, you'll be able to notice a vision that frequency-matches that feeling. What single idea can you focus on that seems like the most interesting next thing, that will assist your growth process? What steps can you take toward it? If you see that what you want is beneficial, it's easier to feel entitled to have it. A new house, for example, might facilitate an experience of a much more focused and expanded sense of self, which will help you contribute more to others.

4. **Notice the idea, pay attention, imagine it, and feel into it.** Grant it life; love it. Keep paying attention and imagining it taking on greater sensory reality. Let the idea drop down through your brain: see it, hear it, touch it, taste it, smell it. Invite it to be part of your life. Feel the experience it's likely to provide.

5. Optimize your personal vibration and connect your personal field with the unified field. When your home frequency becomes your personal vibration, filling your body, emotions, and thoughts, imagine yourself as a tuning fork and "strike your own tone," imagining your clear vibration ringing out through your personal field into the larger unified field beyond you. *Your personal field acts as a set of instructions, or a filter, for the unified field.* The way you treat yourself, for example, is the way the world treats you.

6. Include all beings in your materializing process. Imagine all the souls and particles shifting into place, attuning to your personal vibration, so they can help materialize the next step of your growth, and theirs. Warmly welcome them and express joy and gratitude to them, knowing you're doing the same thing for others. Relax and let the field have time to pattern itself according to your vision.

7. Keep paying attention to the materializing idea. Feel it filling in to the point where your body recognizes it and becomes excited and happy about being kinesthetically engaged with the new reality. Keep investing attention until the new reality occurs and seems like a normal part of your life.

Try This!
Materialize and Dematerialize Three Things

1. List three things—objects, resources, opportunities, or people—that you're ready to have as part of your world. Then list three things you're ready to let go of, that you're not interested in anymore, that you'd like to dematerialize: it could be an addiction, excess weight, or clutter, for example.

2. Take the three items on your materializing wish list one by one: notice the idea, imagine the item, and add sensory detail to your description of the item. Let the item become an experience, let your body get excited about having an experience with the item, and keep investing attention and filling in the item's reality until your body can feel it as normal. Then let go and go on to the next item.

3. When you're finished, imagine how your life feels now that all three items are a normal part of your reality. How have you grown?

4. Next, look at the items on your dematerialization list, one by one. Focus attention on the first one. It's a normal part of your reality. Pull your attention out of it. Stop resisting it and let yourself feel bored and uninterested. Does the item have a symbolic meaning or an emotional tie? Let it fade. Let yourself feel unmotivated concerning it. Let yourself feel appreciation for what it's given you, bless it, and send it on its way. You don't have to take action now, just release and dissolve it in your mind. Then let go.

5. Imagine what your world feels like with these three items gone from your reality, with fresh space there instead. How might your soul expand into the new space?

What Goes into Being Lucky?

Beth entered a home show contest and became one of twelve finalists who had a chance to win a backyard makeover worth ten thousand dollars. She called me, excited about the opportunity to experiment with her awareness. "So how can I be lucky?" she asked. I advised her to feel her attitude and energy as nonchalant and flowing, for her to think the process of winning the makeover was natural, easy, and "normal for someone like me," that "I am a person who has good luck and often wins things. I am naturally open and welcoming. I give generously and receive just as easily." I said she could merge with the house and yard, talk to them about what "we" would like to do to experience improvement and expansion, and ask them to help attract the makeover. And she could imagine appreciating the people who would implement the makeover and feel the makeover finished and taking hold, with the new plants growing happily. So Beth set off to adjust her awareness and energy level.

And she didn't win. I asked her why she thought it hadn't happened, and she said she'd gone to the landscaper's website and seen that they specialized in features she didn't particularly want. She said that of the eight garden designs on the website, she only "kinda sorta" liked one, maybe a "four point five out of ten." She said of the features she did want, the landscaper charged nearly ten thousand dollars for just one, and "I thought we—the garden, trees, home, nature devas, and me—would not be happy with this designer's limited vision."

Some time later, I received an email from Beth titled, *What I Did Manifest!* In it she said, "There is such elegance and humor in how the universe responds to me once I let go and trust that something even better is just around the corner! My husband's ninety-one-year-old aunt, whom I dearly loved, just died. We had taken care of her garden for the last three years because it brought her great pleasure, and she couldn't bend over anymore. I loved doing it for her every week and made sure she always got what she wanted. Now she's doing the same thing for me. Two weeks after the home show drawing, I got a letter from the estate attorney saying we'd been named as beneficiaries in her will. The money is now there to do the yard makeover that I *really* wanted, and what's even better, the garden will remind me of her and our love for each other. I feel amazingly lucky now, like my body is full of champagne bubbles. I'm just so delighted I didn't win that contest!"

Luck may well be that state of consciousness in which you trust the soul, the moment, the perfection of the way your needs are being met, and the timing of events. Your soul may facilitate a lucky experience for you so you can experience the details of the way it feels when things are working optimally, so you can recenter into it at will. In Beth's case, the experience did not let her mind stop short of a truly inspired outcome. You can see yourself as lucky on a day-to-day basis—just look at how opportunities present themselves, how synchronicity occurs to bring attention to certain ideas, or how the flow of each day can become a thing of beauty. By choosing to feel lucky, you are choosing your home frequency and reaffirming the principles of the Intuition Age. And since you're not voting on how grand or dramatic something has to be in order to be termed "luck," you'll find that the quality of what shows up for you will exponentially increase.

Henry, who lives in Nashville, began experimenting with the "pay it forward" concept a few years ago, by handing certain people a $20 bill and challenging them to use half for themselves and pass along the rest to someone else. In every case, he says, the results showed him that when you release energy in any form, it becomes part of what you can draw from. The first time he gave one of his students $20, the young man drove to the Nashville Mission and paid for one night's lodging for a homeless person. The next week, the student received an unexpected check for $50. Henry put a set of tires on a friend's car and in two weeks received a call to do a freelance project that paid three times what he'd spent on the tires. He has come to know that materializing is really a matter of freely giving and receiving.

> We live in an ascending scale when we live happily,
> one thing leading to another in an endless series.
>
> Robert Louis Stevenson

You Can Materialize a New Life Direction

There are times when a habitual way of living or working, or being surrounded by a group of people whose vibration is lower than yours, numbs your awareness and partially hypnotizes you. You know you need to spark things, perhaps by taking your next courageous action, but you can't exactly see what it is, and the timing seems wrong. How to materialize a new direction in life? How to rejuvenate your happy, youthful, optimistic, adventurous self? Should you throw yourself into a swift-moving river? Give up everything and wander off with your begging bowl? Stage a full frontal assault?

Jim has had a fascinating life. He grew up on a farm in Australia, worked as a trainer for the Peace Corps in many countries, was a dance instructor and a videographer, and is now a busy, high-powered trainer for many government agencies in Washington, DC. He reads voraciously about every new development in consciousness and personal growth and is fascinated with merging ideas from the new physics and brain science with leadership skills. He longs to integrate the material into the trainings he does, yet his audiences are locked into the government's narrow perspectives about how to be successful. This leaves Jim modulating his personal vibration to match a lower frequency—not great fun in the long run. We've spoken many times about his need to weave together his various colorful threads and how he might create his own body of material, website, and clientele to match his high vibration. But it's been difficult for him to imagine how he could go cold turkey and leave a lucrative career for one that might not provide the same security.

Recently, he felt drawn to recap his career and write out everything he's done, putting numbers to it all. He's lived and worked in twenty-two countries, designed dozens of training curricula and formats, trained ten thousand people, etc. By making this list, he inadvertently renewed his confidence in himself. Next, he went online to research the author of a book he liked. As he jumped through various links, he discovered a call for papers for an interesting-looking conference that was focusing on the ideas he loves—in Australia. Hmmm. Some good synchronicity here, he thought, and got excited. They want people to talk about the things I'm already doing and my favorite ideas! This is a no-brainer—I have to do it! Then he noticed the deadline was in a few *hours*. No way, he thought. But then he countered, Why not just call them anyway?

He connected with a woman and told her how excited he was that they were doing the conference, how he was working with the same ideas, how he'd love to present a talk but knew he'd missed the deadline. He was pumped. Because he'd just quantified his career, he had his statistics at his fingertips and was able to speak clearly about who he was and what he'd done. He told her how happy he was to have people to talk to about these leading-edge ideas and said that if they decided to extend the deadline, he hoped they'd notify him. She replied warmly, "Consider it extended."

So Jim sat down and outlined a presentation, even visualizing the room set-up. He found a huge body of material crystallizing in his mind, and for the first time in his life, he realized he was enjoying the writing process. The proposal came together effortlessly, and on the heels of that, he began creating his website, wrote an article, and created a video clip. When we talked, he said the trip to Australia had already paid for itself, and if by some remote chance he wasn't chosen to present, he was going anyway. He said he realized he hadn't known how he wanted to show up as his "new self," that there are times when, "You know you can. You don't know how or what, but you know you will," and this had that feel. He joked that he had set himself up to direct his own attention in such a way—through the synchronicities he found so exciting—that his energy would flow in a particular manner and project him into a new context. He'd moved from a place of feeling somewhat trapped and dull into a highly energized new reality in which he could see exactly how probable a variety of new developments were.

Looking at Jim's process, we can learn some key lessons. He instinctively took a small step out of his habitual reality to get an overview of his life. By summarizing his accomplishments, he completed an energetic phase and moved beyond his blind immersion in it. Then he opened himself to noticing what interested him and paid attention to the synchronicities and high-energy information that matched his own personal vibration. He didn't take no for an answer; by calling the people in Australia, he stayed in his home frequency and took the next step into the unknown. He shared himself openly, without expectation, and told the woman what he loved, giving her the chance to help him have it. By staying in his enthusiasm, he overcame a lifelong dread of writing, and as the high-level content for his training emerged, so did a loose understanding of how his new process was likely to unfold. His soul set up a series of experiences for him, and his mind did its job of noticing and choosing to act at the right time.

> Man grows to the width of his intentions, dwindles to the
> contractedness of his intentions. God starts a man but
> the man will have to finish himself out.
>
> Rev. Charles H. Parkhurst

When What You Want Doesn't Happen ...

Wanting something and thinking about it intentionally with positive emotions for a period of time doesn't mean you'll get it; it may not be needed in your soul's plan. Or there may be other reasons connected to underlying frequency principles that prevent an outcome you've envisioned. Mark is a young, highly motivated management consultant who for years has been driving himself toward the goal of becoming a new kind of business expert. Recently he worked with a team of consultants, spending considerable time on an assessment of a large company and a proposal for a reorganization plan. He was sure they'd get the contract, but at the last minute the company said no, despite the fact that stalling was going to cost them more money than hiring Mark's team would. He couldn't figure out why the contract hadn't materialized when it seemed so obvious that it was in the cards.

As we talked, he realized that his high level of enthusiasm and innate understanding of what the company needed were coming from his own fairly sophisticated personal vibration. He was used to living in the present moment and, accordingly, assumed progress could occur quickly. He forgot to factor in the company's rather old-school management style and several highly risk-averse directors. The company's personal vibration was slower than his, and this caused management to lock down in fear of being overwhelmed by what they perceived as too much to do in a short time. The frequency difference in their realities caused the company to see "what's possible" in a totally different way than Mark did. It was almost as if they were occupying two entirely different worlds, and Mark seemed to be ahead of them—living in their future. I sensed that Mark would get the contract—in about four months—after the slower-frequency thought process of the company's directors had had time to run its course. If they'd been able to open up and trust Mark and learn from him, the transition could easily have taken place in an abbreviated period that fit Mark's perceptual reality, and the solution could have been elegant instead of snagged.

There are several other reasons that something you want may not materialize in your life. It may not be the best solution, or the soul may know something about the future of a path you're attracted to right now that will negatively impact your growth. I had friends who were drawn to New Orleans and wanted to retire there. They collected information, studied real estate prices, visited numerous times, and just as they were about to put their plan into action, their successful entrepreneurial son announced that he'd built a guest house for them on his large property near Boulder, Colorado, with a beautiful view of the Flatirons. At the last minute, they swerved and decided to spend a few years with family before they went to Louisiana. Shortly after they moved to Colorado, Hurricane Katrina hit and they were spared a great deal of devastation.

You may be using too much willpower to "try" to materialize something, or you may be using your own life force energy to fuel and control the materializing process. Forcing something is an indication that you're out of harmony with the flow and are missing some key information. It's common for entrepreneurs to loan a project some of their own energy, as though it's venture capital, hoping to recoup it later. The woman I mentioned in chapter 6, who was developing a green real estate project, did this to some extent—partly because she is so capable and generous, and partly because she holds the vision so strongly that it feels as though the project *is a living part of her* and she wants it to come together in a certain sacred way. She extended herself, offered herself, and has experienced some interesting repercussions because she didn't let the results materialize out of universal energy.

The other partners subliminally and energetically experienced her as having "too much ownership" in the project and thus unconsciously vied for position and power, trying to force her out. The process was actually trying to correct itself so the project could be a proper cocreation coming from the evolving, living vision of equal partners. I've often seen this happen with relationships, but the same is true with materialization: if you take three steps toward someone, and they only take one toward you, you'll be rejected by two steps' worth so that the amount of energy being contributed to the relationship stays even. In materializing, if you push toward an end result with willpower or any sort of controlling energy, the field will stall you by the amount you've overextended until it can pattern itself to flow without force. Then the next step naturally occurs.

> You are precisely as big as what you love and precisely as
> small as what you allow to annoy you.
>
> Robert Anton Wilson

You Can Interfere with Materialization in Several Ways

Another reason something might not materialize is that you're motivated by a negative emotion, such as fear or greed. Perhaps you're praying for something because you dread the alternative. "I have to get more clients so I can avoid foreclosure on my home." "I want a relationship so I don't have to be alone and feel scared." You're feeding your most intense attention to the feared idea, not the thing you say you want; therefore, the dreaded option is the one that will materialize. You might temporarily garner enough force to power through and find clients or a relationship, but the reality you're afraid of is still there and will resurface. There's sabotage hidden underneath: by "trying" to avoid or have a reality, you remain stuck in separation, ego, and willpower. Add to this the fact that you're prone to worry and doubt when you're in this vulnerable state; both of these act as short circuits in the flow of attention into the materializing idea, effectively blocking the end result.

Sometimes what you think you want doesn't show up because your body can't feel it as real or the result wouldn't really let you experience deep comfort. Materializing money can be challenging sometimes because it's a bit abstract for the body, since money is often just numbers on paper and not particularly fun or stimulating as a physical experience—plus, physical money is often dirty and smells bad. Bodies tend to quickly materialize things *they're* interested in. Beth's backyard makeover is a good example of an unmotivated body blocking a result. Her body did not like the way it sensed the outcome would feel; the designer's style was a different frequency than Beth's, and her body squirmed at the thought of having to live with something that wasn't nurturing. This principle applies to the learning process as well. If information is presented in a purely mental context, with jargon, it may speed right past you. But if you're given a sensory example, an analogy, something your body can imagine and experience, the ideas make sense in a grounded way and many associations and connections are often made instantly. Without the body experiencing the "reality" of a pattern of ideas, the circuit to meaning and materializing is not truly completed.

Finally, you may not get something because you may not feel entitled to have it or don't really want the changes it would bring. This may originate from an early emotional wound, a karmic vow of poverty, or simply from not stretching yourself enough. If you're used to a certain level of income and what it can afford you in the way of comforts, or if you're comfortable with the way your business functions, you simply may not want more—more might complicate your life. Or, your imagination may have calcified slightly over the years, and if you haven't exercised it enough, you may not be able to dream up bigger, more complex things unless you work to jumpstart it. Part of expanding the dream of what you'd like to create is also dreaming up the wherewithal to be able to accommodate the expanded energy, activity level, knowledge, physical space, and new sense of self.

If you think an expanded life will overwhelm and exhaust you, you'll block it for sure.

Speeding or Stalling the Materializing Process

YOU SPEED MATERIALIZING WHEN:	YOU STALL MATERIALIZING WHEN:
• You notice an idea and invest attention into it.	• You see what you want as separate from you.
• Your body experiences the idea as real and normal.	• You want out of fear; you worry and doubt.
• You radiate your high home frequency through your personal field to the unified field.	• You try to force the outcome or use too much of your own energy to make things happen.
• You honor all the other beings and particles that are cocreating with you.	• You don't feel entitled, or you lack imagination.
• Your personal vibration frequency-matches with the item, people, situation, place, or experience you're seeking.	• What you want is not on purpose for your soul, or your soul knows about a future impact of the result that would be harmful to you.
• You trust the outcome will be perfect.	• Your mind tries to do more than its share of the work.

What Does Success Look Like in a High-Frequency Life?

Success in the Intuition Age starts with this: If you see your world as high-frequency, it will cause you to be high-frequency. If you see yourself as high-frequency, your world will be that way, too. Before transformation, when something we want doesn't materialize, we're so quick to say, "What's wrong with me?" After transformation, we say, "Which part of the process am I forgetting to honor?" In the Intuition Age, you're motivated to live your destiny, which helps you choose to materialize what's in alignment with the most natural and graceful expression of your soul. You're clearly aware of your affect on others and theirs on you, and you work with that principle compassionately so everyone benefits.

You understand that there are no limits to what can be provided, that anything can materialize, yet you also know you're built to do certain things, that you have particular life lessons, and you'll receive exactly what you need at the right time and in the right form. You may be drawn to accomplish goals, but it will be for the fun of it rather than to solidify a sense of self-worth. You may find that it's fun to dance your way through life, responding to the waves of materialization and dematerialization, to the mysterious urges that appear in your awareness, to the sudden changes of direction, to the ever-evolving experience of the diversity and depths contained in your Self. When you don't hold limited ideas of what you are and what life can be about, you are much more able to shape-shift—to respond instantly to whatever life needs you to become.

All physical beings have communication from their inner being in the form of emotion, and so, whenever your emotion is positive, you can know that you are in harmony with your inner intention.

Abraham/Esther Hicks

You engage with what comes. A thought about creating an event may present itself, or someone may call, inviting you to do some work. It doesn't matter whether the source is internal or external; the ideas all come from the soul, from the Us. You understand the *law of correlation*, which states that because the inside and outside aren't separate, if a thought occurs to you, it must also be occurring—or soon will—as an event in your world. And conversely, if a dramatic event happens to you, you know you hold a corresponding idea that brought the event into your awareness. By connecting the inside and outside, you become skilled at "reading your reality" for clues about what's going on inside you and you pay attention to how what you think patterns your reality. Everything educates you.

The more you embody your soul, the more instantaneous materialization becomes. When you need help, an expert appears. You learn to love the synchro-mesh way that life functions when you don't get in the way with immobilizing thoughts. You are fascinated with honing the skills of living in the moment and sharing responsibility for the direction of your life with the collective consciousness—which arises from each particle of light, in each moment.

Survival of the Fittest? The End Justifies the Means?

Compare this, then, to our current ideas about success: Survival of the fittest. Those with the most toys win. The end justifies the means. These old ideas are about power over life, not being one with life. To gain an "advantage," we think we must be better than others, and it's become fashionable to be an expert in sarcasm and put-downs, to attack others in clever, mean ways. The righteous expression of ego is projected like a weapon, and audacity wins over wisdom. People are impatient and feel they must *take what is rightfully theirs* rather than allow the universe to give to them freely through their home frequency. They don't realize that it actually takes less energy to work with the soul than against it—and no one needs an advantage because we're all born privileged, with equal access to frequency principles.

Mark, the young management consultant I mentioned earlier, told me about two older, extremely brilliant colleagues who had made an art form of interrogating their clients with "zinger-gotcha" questions that seem smart on the surface but are actually meant to show the men's superiority and reduce the client's self-esteem. Mark noticed that the clients were becoming intractable and was himself uncomfortable in the room when this was going on. He had thought that because the two men's energy was fast and aggressive, they were high-frequency. In actuality, their frequency was low because they did not feel how interconnected they were with others. They were emotionally wounded, and in being so smart and aggressive, they perpetuated that pain by wounding others. This simply gets in the way of the field operating elegantly and materializing perfect results.

We are tempted today, with so many challenges in the financial realms, to choose partners and colleagues who have gained material abundance via selfishness, willpower, and dominance. Who cares how they got it, right? Yet anyone who uses methods that come from the old linear model of perception based on separation—they can't feel or don't care how their actions affect others—have unattended-to, unhealthy feeling habits. That means their personal vibration is low, and no matter how dynamic they seem or how much money they bring to the table, in the not-too-distant future their internal time bombs will sabotage their results. If you're involved with these people, you'll go down with their ship. The reason is that willpower and charm can only jack things up for so long; then the hidden underlying misperceptions surface to be cleared, and the reality realigns to match the unhealthy feeling habits. At that point, such people experience loss—until their lessons are learned and transformation occurs. Today, with the accelerated frequencies in the earth and the rapid clearing of the collective subconscious, these collapses are happening faster and more often.

Yet we must remember that while some people go through stage 5 of the transformation process (dissolution of old forms), many people are simultaneously becoming clear at stage 6 and beyond. The world is not going down the drain; there is balance. All you have to do is choose to vibrate in harmony with your soul's inner peace and love, and if you get confused, choose to do it again. There's no need for fear. Look beyond surface appearances—what is this person or opportunity really like at the core? Is there much difference between the book and its cover? What sort of people do you want to spend your precious time with? Validate the way people create—the means—as much as what they have, the ends. In the Intuition Age, process and experience are more important than results.

> The important thing is this: To be able at any moment to sacrifice
> what we are for what we could become.
>
> Charles DuBois

Materializing financial resources in the Intuition Age is always going to be a matter of seeing first what's most interesting and joyful to your soul. Then you let yourself feel naturally motivated from your body's point of view, thinking of the *experience* as something that benefits you and others. You allow money to flow in and out as a symbol that measures how much you want or need to create, and you see finances as oil that helps grease the wheels of your energy flow. There is plenty of flow everywhere, and many experiences can be had for free. Keep your eyes on the fun experience you want, and the money, if needed, will be drawn in from the periphery to support your forward movement. In the Intuition Age, not having enough and having too much are seen as disservices to your fellow souls, as things that act as subtle interferences to an easy flow.

Imagine How Your Life Might Shift

So what might your high-frequency life—your destiny—look like? Scan through the circumstances of your present life: relationships, work, recreation, health, satisfaction levels, and the leading edges of your spiritual growth. Imagine your life as waves and frequencies of energy; there are places you're trying to control, places where the motion is flowing or paused of its own accord, and places where the frequency is naturally high or low and contracted. You'll find unhealthy feeling habits beneath the places that feel dissonant or that you're trying to control.

Once you assess the different areas and how they're working, add extra energy to each area. Imagine flooding each area with an unlimited amount of love so you feel unconditionally loved, lovable, and loving. You feel safe and relaxed, then happy and curious, then adventurous and creative. The power of love and innocence fills up, expands, and rejuvenates the areas outlined above and transforms them as suggested below, specifically:

- **Notice places where you feel there are problems or contractions**, where you're holding too tightly to something, avoiding something, or where other people are affecting your decisions in a restrictive way. As energy is added and these areas loosen and move, you receive insights and learn more about yourself.

- **Notice places where you're bored, where things seem old or dull**, where you're going through the motions or feel restless. What are you still making yourself do? In what areas do you feel exhausted and drained? As energy is added to these areas, you find and receive the benefit from what's been given and let the situations dematerialize. In the resulting open space, you can rest and realign with your soul.

- **Notice places where there is easy flow**, where you look forward to involvement, where there is a pooling of your home frequency energy. As energy is added to these areas, they amplify and often make a "turn" or jump an octave into even better expressions of soul.

- **Notice places where you want to step forward into the unknown**, go beyond your comfort zone, or take your next courageous act. As these areas receive more energy, hesitations dissolve, and it's easy and exciting to stretch to be a more talented and competent person. You receive a "reality sense" about a new direction.

If you take time to go into each area, see how unlimited love affects it, then put all the insights together, you'll catch a glimpse of what your destiny, and your high-frequency life, can be.

Try This!
Zero In on Your High-Frequency Life

Relax and let your soul, working through your imagination, give you some impressions about what's possible in response to the following:

- What are three surprise directions your life might take?
- What are three surprise strokes of good luck that might come to you?
- What are three surprise locations you might be connected with?
- Who are three surprise benefactors who might assist you generously?
- What are three courageous acts you might take next that would make you proud of yourself?

Just to Recap . . .

When using energy and your personal vibration to help materialize and dematerialize things in your life, use holographic models based on unity rather than linear models based on separation, because this makes the process faster and more accurate. Ideas materialize not because you attract them, but because you notice them already inside your field and invest attention in them until they become real to your body and normal. The soul is responsible for the first level of materializing your important growth lessons and circumstances, and your feelings and mind are secondary, serving to either block or assist the soul. You can intentionally focus on things you want to create, and they will materialize if they serve the soul's purposes. There are a variety of frequency-based reasons that things do not materialize when you want them to, such as using too much willpower, not feeling the result as real in your body, or the soul knowing something about a negative future impact of the result that wouldn't be on purpose for you.

You do not create alone but as a function of the cooperation of particles, waves, and other souls. Luck is a state of awareness in which you trust the soul, the moment, the perfection of the way your needs are being met, and the timing of events. In the Intuition Age, we need to see success as being more than material wealth and superiority over others and be cautious about who we partner with, as the hidden time bombs in people who have attained success through methods based on a belief in separation are likely to explode and cause failures soon. Your destiny is your high-frequency life, and you can help materialize it by clearing unhealthy feeling habits, choosing what's choosing you, and engaging fully with what you already have.

Home Frequency Message

As I explain on page xxi in *To the Reader*, I've included these pieces of inspired writing at the end of each chapter as a way for you to shift from your normal, speedier reading mind to a deeper kind of direct experience. Through these messages, you can intentionally change your personal vibration.

The following message is meant to transport you into a way of knowing the world that's close to the way you'll experience life in the Intuition Age. To move into the *home frequency message*, just downshift to a slower, less hurried pace. Take a slow breath in, then out, and be as calm and still as possible. Let your mind be soft and receptive. Open your intuition and prepare to *feel into* the language. See if you can experience the deeper realities and feeling states that come alive *as you read*.

Your experience may take on greater dimension in direct proportion to the amount of attention you invest in the phrases. Focus on the words a few at a time, pause at the punctuation marks, and "be with" the intelligence delivering the message—live and right now—to you. You might speak the words aloud, or close your eyes and have someone else read them to you and see what effect they have on you.

GROW IN INNOCENCE

Innocence is one of the least understood qualities and powers of your soul. Far from denoting emptiness, helplessness, or passivity, it is a mighty unifying and activating force. Like the Midas Touch, innocence catalyzes wisdom and experience of trust, abundance, and perfect provision wherever it penetrates. Therefore, to create your best life, grow in innocence.

Look in the baby's eyes: here is soul light undampened by shadows, openness with no strain of vigilance. In the shine: availability, willingness to engage and play. In the baby's gaze: innocence that when entered as a shared space brings you back to ancient wisdom. Travel into the space where the baby's gaze originates, feel what motivates that light-that-loves-to-connect, feel yourself as innocent again. As you do, your frequency rises.

In innocence you are ready for anything, welcome everyone, trust unconditionally. You do not hold back but respond spontaneously, and respond again, for life is ever new. Whatever you notice is a gift, it is yours. When you stop noticing it, you're noticing something else. One reality fades as another arises. Loss does not exist. Things come, as if by magic. Imagine how it feels to double or quadruple your innocence. In this intensity, you beam even more because now you know to expect what is best, what takes care of you. You thrill at the surprise of the gifts and pleasures coming freely to meet your needs and be with you.

Imagine intensifying your innocence even more, and you feel how many concepts dissolve, like scale, need, and even self-worth. Now there is just interest, just delight. Here you and the earth become playmates, sharing energy and loving each other as siblings. Life proceeds in you both, unlimited and full. You are speed-of-light kaleidoscopes; millions of ideas, forms, and experiences are materializing and dematerializing through you, and you are becoming more conscious and entertained with each creative cycle.

Do you wish to materialize something? First be like the babe, and soften your eyes, mind, heart, and body. Shine out! Smile and sweeten your perception with ready-to-be-amused innocence. Think of your desire for a new experience and, in your timeless innocence, expect a perfect form to surprise and please you. You and the Earth, together in magnified innocence, expecting the creation wave, any minute, to arrive and tickle you . . . Here it is! And now, shall you go again?

10

Accelerating Toward Transparency

The [Duino] Elegies show us . . . the continual conversion of the
precious visible and tangible into the invisible vibration . . .
into the vibrational-spheres of the universe.
For since the various materials in the cosmos are only different
vibrational-rates, we are preparing in this way, not only intensities
of a spiritual kind, but—who knows?—new bodies,
metals, nebulae, and constellations.

Rainer Maria Rilke, from a letter to Witold Hulewicz, 1925,
translated from German by Rod McDaniel

As you reach the last phases of the transformation process, you've passed through the eye of the needle and it dawns on you how far you've come. It's awesome, really. You've shifted from thinking of yourself as just a solid physical body with thoughts and emotions to experiencing yourself as a series of interpenetrating fields of awareness, all vibrating at octaves of the same home frequency—which is the tone of your soul. You've learned to stop identifying yourself as a wounded and self-sacrificing person and to clear blocks in your emotional and mental realms. By dissolving your shadows and not holding back the waves of energy that come and go through you, you've begun uncovering the you who's always been inside, the you made of diamond light and love.

The more transparent you become, the more your soul freely expresses through you without distortion to create your destiny. Problems dissolve and become simple shifts of thought and direction. Now you're used to the way your thoughts and feelings instantly affect your world's form, and you know what to give attention to and what to ignore to help your soul shape your reality. What used to be difficult is now interesting and entertaining. As you let yourself receive and give without restriction, responding to life's surprises with an open heart, your identity becomes less fixed and more fluid. You experience a wider variety of options and feel more expansive and free to dream new dreams.

It seems matter-of-fact that we live in one expandable-contractible present moment, that all beings, forms, and fields are conscious, compassionate, and cooperative. The Golden Rule—treating others as you would like to be treated, or as the Divine treats us all—seems so *obviously logical!* You're sensitive enough now to understand how even subtle negative thoughts can interfere with the inspired flow of life, and you can't imagine doing even this tiny amount of damage. Perhaps you've never thought it possible, but you're nearing the point of self-realization, or enlightenment. In fact, if you know how ecstatic you can feel just resting in your home frequency, enjoying the simple pleasure of *being*, you're very close, indeed.

The path up is the path down. The way forward is the way back.
The universe inside is outside but the universe outside is inside.

Robert Anton Wilson

Spiritual Growth Unfolds in Staggered Waves

Still, you're in a process that takes time. You have breakthroughs, moments of crystal clarity, and deep experiences of love, but in short order you can become lost in the collective confusion again. Then you remember what's really real and how you prefer to feel and recenter in your truth. As long as you live in a body, you'll oscillate like this, but it becomes easier to shift back to your home frequency and second nature to live as your soul.

There is great compassion in the evolution process: the moment you take one step toward transformation, then another, even the most difficult tasks are eased. What Paolo Coelho says in *The Alchemist* is true, that when anyone is truly trying to live their destiny—when they ask for something with this intention—"all the universe conspires to help you achieve it." While all you know may appear to be breaking apart from the launch velocity of your rocketing life, inside, your loosened particles are quietly emitting diamond light that wafts gently through you and out into space, stabilizing and saturating you with the wisdom of your soul. Inside, you are softer and less contracted. Something in you is warming, opening, and becoming spacious.

The transformation process doesn't happen for everyone at the same time. It unfolds in successive waves or ripples. Some people become clear first. They're no better than anyone else—they've simply agreed to go first. These people influence other people, facilitating more clarity by their own example. As the second group passes through the process, they attract another ring of people, who attract another. There's truth in the saying, "The first will be last, and the last will be first," because in becoming clear, we discover that we all reunite with the Divine *together*, since we are *one* collective consciousness. That means that when you find freedom, you naturally choose to help others who are still stuck and confused. The soul has no other wish than to have an unlimited number of fearless, joyful playmates on the journey home.

> Ours is not the task of fixing the entire world all at once, but of stretching out to mend the part of the world that is within our reach. Any small, calm thing that one soul can do to help another soul, to assist some portion of this poor suffering world, will help immensely. It is not given to us to know which acts or by whom will cause the critical mass to tip toward an enduring good.
>
> Clarissa Pinkola Estés

I believe we'll see a time when many old structures collapse—*bam! bam! bam!*—close on the heels of each other, and millions of people will enter the letting go and "just being" period together. At the same time, there will be an increase in the numbers of enlightened people whose lives are stable and whose creativity is open and active. The critical mass of these consciously ensouled people will have the power to catalyze the self-realization of a vast number of the others. Amid the dissolution of the outmoded will appear miraculous innovations and developments, growing like seedlings through the cracks in the rubble. It promises to be dramatic and exciting, but meanwhile, back in *this* now-moment, becoming clear is not that lofty or glamorous; we must just stay present, keep on walking, and continue choosing the home-frequency feeling of the soul's reality, building an unshakable habit of being in tune with evolution's loving process.

It's Okay to Lighten Up and Laugh About It All

I recently spoke to Liv, a social worker, about the negativity and fear so many people are feeling. She told me the people who work with her were using the excuse of the latest terrible blizzard to act out their feelings of panic and victimhood. As someone who deals with real pain and suffering every day, she felt something might be wrong with her because all she wanted to do was laugh. Every sob story amused her and she was embarrassed by her "perverse" response. She had to retreat into her office to make sure she didn't offend someone. I had experienced something similar—while the news was becoming more grim, I was more cheerful. My life felt light as others were paralyzed for weeks by the flu, forced to move or change jobs, or were going through difficult divorces. How is this possible? How can we feel clearer and happier just as the world seems to be entering the densest, most hopeless part of the transformation process? Are we callous and insensitive, or just plain loony?

These feelings are not uncommon lately among people who've been on a spiritual path and working with transformation for a while. Lightening up doesn't mean that compassion disappears; in fact, it increases dramatically. It does mean that you know the importance of dissolving negative thinking, and as you embody your soul, it's much easier to see other people when they're caught in the ego's self-important dramas and victim consciousness. The two realities—clarity and confusion—become much more discernable as the gap between them widens. You can see and feel others' pain, and you can also see how their minds hold the pain as meaningful or special instead of identifying with their beautiful souls. Sometimes when you see it from just the right perspective, the whole human drama in relation to the true scope of the soul seems downright hilarious, like some cosmic joke. Perhaps this is the way we release the tension of having defined our human journey as so terrible for so long and return to the simplicity of being.

If you find moments of true laughter (not at life, but with life), it's definitely good to take advantage of them and shine out to those around you. It's definitely good to stay in the truth and invite others into it with you. You no longer have to feel guilty for feeling clear and good. You don't have to distance yourself from people who are suffering—no matter how devastating and cruel it appears. The greatest gifts you can give those who are suffering are your open heart and warm attention, and your real knowledge about how to shift realities.

Just What *Is* Enlightenment, Anyway?

Looking at the world, so full of negative emotion and low vibration, it's hard to imagine that we could be nearing a state of enlightened awareness. If you've fallen back into frustration, irritability, exhaustion, or depression—or if your main focus is on family, health, or career—the whole concept of self-realization or enlightenment may seem ridiculous, "out there," or way in the future. It may be the last thing on your mind, but I assure you, it is the first concern of your heart and soul. When your defenses are down from overexertion and your mind is not in control, you might just shift into the next level of self: a state of transparent awareness in which nothing interferes with your soul expression.

The great Buddhist teacher Dogen said, "Do not think you will necessarily be aware of your own enlightenment." Perhaps that's because the enlightened state isn't considered to be an endpoint but an ongoing path where the goal recedes and expands as you approach it. Enlightenment is a word that throughout history, especially in the West, has had to be used sparingly for fear of persecution for charlatanism or heresy. In the West, there is the concept of *salvation*, in the East there is *nirvana*, *satori*, or *mukti/moksha*, and in Islam the Qur'an speaks of a high state called *peaceful soul*. I suspect that not many of us have a clear idea of what these terms mean or what the state of spiritual—not just mental—enlightenment might actually *feel* like. How might you change when you get used to living in your body at the highest frequencies? Mystics of all faiths have probably come closest to the experience through direct merger with the Divine.

Enlightenment or self-realization is often defined as "waking up" to your inner divinity and having a perception of your authentic self and the "true nature." Your worldly personality-self and the experience of duality fade, so you directly recognize yourself as pure consciousness. You experience a peace of mind that is free from craving, anger, and other "afflictive" states that cause suffering and know union with all that is, or with a loving, eternal Godhead. In an enlightened state, the Mind, with all its jerky seizing up and tendency toward conflict, is still there, but you don't identify with it: you watch it "acting out" instead. William Blake packed a lot of meaning in his simple description of enlightenment, saying that it is "taking full responsibility for your life."

Some of us may confuse enlightenment with *ascension*, where the body rises in frequency so fast that it disappears from physical reality. In the West, it is said of Enoch, great-grandfather of Noah, that he "walked with God, and was not, for God took him." Similarly, Hercules was raised to heaven and made a god by Zeus. Christian doctrine holds that Jesus bodily ascended to heaven forty days after his resurrection. In the East, there are "ascended masters"—Theosophy calls them the "mahatmas"—who appear and disappear at will throughout time in various guises. Perhaps, as many of us become enlightened or self-realized—what I am terming *transparent*, or made completely of diamond light—we'll discover we have the ability to ascend and descend through the realms, taking our physical bodies with us if we want. As we thoroughly grasp the principles in materializing and dematerializing physical reality, it may not be much of a stretch at all to "ascend."

> When I look inside and see that I am nothing, that's wisdom.
> When I look outside and see that I am everything, that's love.
> Between these two my life turns.
>
> Sri Nisargadatta

How Might Enlightenment Feel to Us "Regular Folks"?

I asked a well-educated friend whether he wanted to be enlightened, and to my surprise, he danced noncommitally around the idea. He seemed to see the concept as too presumptuous and final, as though being enlightened might bring such a great responsibility that he might not be equal to it. So I asked him what enlightenment meant to him and what it might be like to live *after* he was enlightened? He laughed and said that it probably wouldn't be much like *his* life and that we don't have many models for it because people who are enlightened are supposed to live in caves meditating to hold the world together—or they pass right by us on the street, so we never see them in action. He said people think enlightenment is a peak experience in which we're engulfed by white light, hear angel choruses, receive the entire collective wisdom of the universe in one blast, and become saints forever after. "And then," he joked, "who would be your friend?"

"But what happens the day after enlightenment?" I persisted. My friend thought about it and said: "I'd probably make many of the same choices I make now, but I wouldn't worry. I wouldn't doubt my abilities or my guidance or my creative urges. I'd naturally gravitate to the loving view and the win-win solution instead of choosing punishment or sacrifice. I wouldn't feel the world is unfair; I'd understand the hidden meaning and benefit of events, and I'd live a simple life. Everything would just be simpler." How true. Not so different, but very different. Imagining enlightenment is a bit like imagining yourself to be the richest person on earth or someone who is immortal. It seems a big stretch from your everyday mindset, yet ironically, it may not be that much different—you'll just know how unlimited you are.

I am convinced that in spite of the appalling negativity in the world, the experience of transparency or spiritual clarity will become common; more people moving through the transformation process will make it easier and easier for everyone. Living in and resting in your home frequency may sound like a shallow thing, but the experience of actually changing your frequency to this unifying tone is all-powerful, and it can transport you into the state of heaven on earth.

There Are Yang and Yin Paths to Enlightenment

I've long had an understanding about achieving spiritual growth that differs from the major religions of the world. Perhaps I'm aware of this because the Intuition Age is bringing a new balance between yang-masculine and yin-feminine energy and awareness. Historically, I think the world has held an unquestioned view of spiritual growth that is based largely on men's experience of how it works for their physiological makeup and brain chemistry. I think women have a different experience of the process. I don't say this out of any sort of feminism or divisiveness but because I sense that a more comprehensive view is opening to us, and acknowledging how women's bodies and brains understand the universe and enlightenment will help round out a bigger picture for everyone. By combining the yang and yin views, we can find a unified way through which all people, no matter their sex, can attain the highest levels of awareness.

In a nutshell, because men's brains have fewer connecting fibers between the left and right hemispheres, they tend to perceive the world in an either-or, one-side-of-the-brain-at-a-time sort of way, which gives them natural skill with analysis and compartmentalization. Right-brain intuition is a mode they must often intentionally activate. Therefore, it makes sense that men naturally understand the Divine through separation; their path is to be in the world and not of it. Men's self-realization is typically based on abstinence from physical "temptations," contemplation and study, structured physical ceremony, surrender of personal will to a teacher, and monastic isolation. With Zen, for instance, all things of the world are treated neutrally as "nothing special/everything special." Heaven, or the Pure Land, is a goal that is "up and out," beyond this world.

Women, on the other hand, have many connecting fibers between the two sides of the brain, giving them the ability to perceive in a both-and way in which separation from others and the world—even between thought, emotion, and spirit—is inherently difficult. Women thrive on relationship, conversation, nurturing, merging, feeling, and intuition. It makes sense that women's path to enlightenment is "down and into" the world, through matter, into everything human. Women tend to know the worlds, both physical and nonphysical, as part of their body. They are about *being* the world. Life is about tending and caretaking, since it's all too easy to feel another's pain. For women, enlightenment isn't a separate goal; it's a place where they come from, deep down.

In the Intuition Age, it seems to me that humanity as a whole is finally bringing its unique kind of perception into full bloom. As human beings, we have a special potential because we're endowed with conscious sensitivity and free will—we can develop our capacity to feel to the nth degree until feeling turns into the enlightened emotions of empathy and compassion, and we can feel our way right into a merger with our Source. Because feeling and sensitivity are body related, our perception of the invisible realms can be more real and personal. That means we can complete our "evolutionary experiment" of being self-realized individuals by consciously experiencing ourselves as miniholograms, or microcosms, of the Divine.

When you are sensitive, you relate to, then include, then integrate everything in the world, much as you eat food. As your experiences digest, you receive the divine food value in it all. As you take in both light and dark, each new part teaches and expands you, becoming part of the diamond light in your personal field. Knowledge becomes personal, complexity simplifies, dissonance harmonizes, and you experience kinship with all forms of life. Differing worldviews merge to form a larger truth. Eventually you become so inclusive you let go of defining yourself. The Divine becomes you and you become it. You know your highest identity by receiving everything; you're holy in all your humanness.

> Humanity and divinity will be identical
> when we recognize divinity in humanity.
>
> Ernest Holmes

Both the yang way and the yin way to enlightenment work, but there is one path that works equally well for all people, and that's the path of the Heart. Where brains and hormones may differ, all our hearts function the same way. Being openhearted and heart-centered means you're in your home frequency, which lets you know the world via feeling and also brings insights that are clear and wise. Your heart is like a doorway to oneness.

Try This!
What More Can You Include in Your World?

1. For just a minute, think about unpalatable things, things you don't like to think about. List half a dozen of them. Notice the negative charge you have concerning each, and take the resistance out of the ideas and

the images. Let each one be a natural part of the world, and of your world. Let each thing be available as an optional reality for whoever needs to experience it. Then love it and be kind to it.

2. Now for a minute, think about things that seem way beyond your grasp, realities that seem far out or nearly incomprehensible. List half a dozen of them. Notice the "unavailability charge" you have on each and take the resistance out of the ideas and images. Let each one be part of the world, and of your world. Let each thing be available as an optional reality for whoever needs to experience it. Then love it and be kind to it.

3. Feel the new realities you included in your world, how they might be useful, and how they are simply experiences that eternal souls might be curious to feel. By allowing these things to be part of you, feel how much more relaxed and expansive you are and how much more of yourself you have access to because you're not caught in resistance.

You Discover an Expanded Understanding of Empathy and Heart

A natural result of reaching the end phases of transformation is that a refined level of empathy—an *empathic resonance*—becomes your preferred mode of perception. When you're in your home frequency, feeling into the world with conscious sensitivity and no hidden agendas, you become highly empathic. Empathy is so much more than is conveyed in the typical definition: feeling someone else's pain. It brings information about what it's like to *be* the person who's just lost her husband or won the lottery, or to be that unmowed yard or that beautiful ripe peach ready to be picked. Empathy is direct knowing through vibration; it is sensitivity filtered through your heart.

As you evolve, the heart literally becomes your new brain center. The Institute of HeartMath has found that the heart's electromagnetic field may play an important role in communicating physiological, psychological, and social information between people. Its experiments found that one person's brain waves can synchronize to another person's heartbeat. Researchers inferred that our nervous system acts as an antenna that responds to the electromagnetic fields produced by the hearts of others. Interestingly, they also found data that point to the fact that the heart's electromagnetic field is involved in intuitive perception, that both the heart and brain receive and respond to information about a future event before the event actually happens. The heart appears to receive the intuitive information before the brain does. This is exciting scientific information that begins to match what we already know experientially.

Empathy helps you avoid the life-damaging choices that make your heart contract. For example, it hurts you and others when you cut someone out of your life because you're mad at them, or justify a bad habit because you're afraid to face a memory, or refuse to try new things because you're self-conscious. Empathy alerts you to these clenchings and reopens the energy flow. Empathy is also a great healing force because it helps you see past the definition of yourself as an isolated, fragile individual. It penetrates right into pain, and through love, vaporizes it, dissolving separation.

> When you love somebody, your eyelashes go up and down
> and little stars come out of you.
>
> Karen, age seven

Empathy may well be our greatest human capacity, an inborn ability to feel the soul within all things. It teaches you about others and the world by heightening personal connectedness. When used regularly, it ushers you into deep and lasting compassion. The next time you have to deal with a faceless customer service representative on the phone, try feeling what their day has been like. Then talk to them as though you really do understand them. See how much better you both feel afterward. Or the next time you're with a friend or loved one, see what happens when you both intentionally focus on being empathic toward each other at the same time. It's an amazing experience that brings great appreciation and a feeling of cherishing each other.

Try This!
Focus on Your Heart

1. This exercise will help you learn to alter your personal field through the power of focus. Wherever you place your attention, there is an immediate flow of subtle energy to that point. Put attention on your

physical heart and notice the flow of energy into it. Connect the feeling of gratitude with your attention; feel appreciation as you focus on your heart. You might feel gratitude for your heart doing its job, for your body, and for many specific things and people in your life. As you scan and appreciate, also keep your focus on your heart.

2. Let the feeling of gratitude intensify. What amazing gifts you have been given! Let the feeling intensify even more until it reaches near-ecstasy, until you feel that you cannot contain it. Let it overflow and pour out through your personal field. Stay with it and be in it. Stay focused on your heart. Let it continue to intensify and brighten. You're establishing a harmonic pattern that resonates to the frequency of accelerated evolution. Doing this often will raise your personal vibration immeasurably.

It is the empathy of the universe moving through your soul that allows your deepest needs to be met, perfectly, in every situation. This was demonstrated to me not long ago when I spoke to a client who wanted a consultation and aggressively insisted that I speak with her immediately, badgering me to rearrange my schedule to accommodate her that afternoon—which I did. She called back an hour later to cancel, saying bluntly, "I don't have any questions right now, so this probably isn't the best time for me to have a session." I said okay, shook my head, but instead of feeling irritated, I reminded myself that she must feel quite troubled to be so insensitive, that things always work for the best and she must know what she needed—or would, eventually.

Not twenty minutes later, another client called—someone I've worked with for years. She had just come out of a meditation and intuitively felt she should ask me what books I'd been reading as possible recommendations for her book club. When my choices didn't fit for her group, I humorously told her the tale of the pushy woman who canceled abruptly, and she said, "Well, why don't I have a reading? Yes, that's what I'll do. *I don't really have any questions right now, but that's probably the perfect time for me to have a session!*" These were nearly the exact same words that had been used only minutes earlier to justify a completely opposite reality! Life was showing me the power of empathy. Because I had opened my heart and been kind to the first woman, the universe showed kindness and generosity to me in a highlighted way I couldn't miss.

Compassion Is the New Evolutionary Force

As I see it, compassion is slightly different from empathy: empathy is related to sensitivity and feeling oneness, while compassion is a broader, more abstract understanding that knows love as the core of every being and situation. I believe we'll see both sustained empathy and compassion actually change the makeup of our bodies, strengthen our immune systems against progressively nasty environmental toxins, and perhaps even change our DNA so we aren't so susceptible to disease and aging. It won't be long, I predict, before we see compassion surpass competition as the earth's main evolutionary method. Survival is no longer based on the archaic principle of the fittest destroying the weakest but on our understanding the hearts of our neighbors on this, our spaceship earth. Living a compassionate life is *necessary* for the world to evolve. This kind of perception lets us perceive our similarities—especially how our differences and pains are similar—and this will help us end war, prevent environmental devastation, and discover our true nature.

Dr. Bruce Lipton, a microbiologist, says that the heart of a cell is not its nucleus but its outer walls. Receptors in the walls allow it to interact with the environment and, according to Lipton, when a cell perceives the environment to be supportive, the receptors open to feed the cell. When it perceives danger, it closes the receptors. Cells can only be in one mode or the other—growth or protection. The logical extension of this is that by living in fear, you cause a state of general closure and nongrowth, or blocked evolution, in your own system. By cultivating love, empathy, and compassion, you move toward transformation and transparency.

Studies also suggest that the DNA molecule is surrounded by an energy field some scientists call the "phantom or shadow energy" of DNA; interestingly, it remains in place for more than a month after the DNA is removed from the body. The scientists speculate that this energy exists *before the DNA is formed* and actually creates the DNA. What might it be? Could it be part of your inner blueprint, your personal vibration, or your soul's home frequency? HeartMath research shows that DNA responds to intense projected emotions by changing its shape. Gratitude and love caused DNA strands to relax, unwind, and lengthen. Anger and fear caused the DNA to tighten, become shorter, and switch off many of the DNA codes. So, if it responds to compassion, might not your DNA evolve beyond disease and aging as we know it?

Try This!
Flow Compassion and Love Through Yourself

In various circumstances over the next few days, think of yourself as a giver and receiver of love. Look around and be a neutral witness to the basic love energy that shapes you and the world around you. See and feel it everywhere. Be accepting, forgiving, and unconditional in your lovingkindness for a few minutes. Breathe love in through all your cells and let it pass through you, touching every part and lighting you up, then send the love on its way, giving it generously to others. In difficult circumstances, ask, "Where is the love right now?"

Knowing Others as Souls and Finding Your Soul Group

Being empathic helps you know people more deeply. Just by putting your attention on someone and being with them, you'll feel their vulnerable places, how and what they avoid, what they need, and how they might heal. You'll understand their deepest motivations, life lessons, talents, and core goodness. Just yesterday I looked in the eyes of a very old woman slowly getting out of a restaurant booth with her walker and instantly saw a movie in reverse of how she'd looked at every age all the way back to infancy. In that few seconds, I felt love for her. It's easy to feel frustrated and irritated because people so often choose to be afraid, remain ignorant, and not see themselves—and when they do that, they don't offer you their best stuff. When this happens, you need to make a small shift of perspective into the highest way you can see them and "act as if" they are that way. Most often you'll find that people live up to your vision of them.

The more you honor people as souls, the more you'll be able to hold differences as interesting and valuable. You'll feel connected to people you never thought you'd like. As you find similarities to more people, you'll understand the deeply cooperative nature of souls working together in both visible and invisible groups. We all belong to a group of souls who are parallel to us in development—a *soul group*—which in physics' terms is really a resonant field of awareness based on a particular energy frequency. A soul group, then, is a cluster of beings who have evolved to a common frequency, which means they often have matching philosophies, knowledge levels, and motivation. These like-minded people may look like siblings or have similar upbringings, interests, life transitions, goals, or even names. They may be friends, family, colleagues, or nonphysical beings in higher dimensions. When you meet them you feel profound relief or excitement; you feel that you already know them and naturally *want to like them* no matter what.

Globally and culturally we are undergoing an initiation. We are moving out of the journey of the hero and the heroine . . . that process of individuation. An archetypal shift or initiation is occurring where we are moving into the journey of partnership or the journey of the tribe. The journey of partnership . . . requires the spirit of cooperation and collaboration [and] requires that we learn about collective leadership and collective wisdom.

Angeles Arrien

Many people feel that their family of origin is not their "real" family. They sense intuitively that they have another family—a spiritual one—and they're always muttering under their breath, "Where are my people?" Have you encountered people you'd swear you already know, who share your ideas and a similar life quest? You and the people in your soul group are so attuned to the same home frequency that it's easy to feel you're soul mates, soul siblings, and part of a spiritual family. Just contemplating the idea that you belong to a spiritual family or soul group can bring great solace. It helps to know you're not alone, that there are wise people who know you and are ready to assist you. By asking to know who's in your soul group, you'll begin to notice people who fit the bill—both in this world and in your dreams.

When you imagine your soul group, no matter whether you know them or not, no matter where they are—they are immediately imagining *you*. In the unified field, there is no chicken-or-egg dilemma; attention is simultaneous and flows both ways. You create each other at the same time, and you create your mutual experience equally. If something is initiated, it's caused by all participants. By imagining your soul group, you can have helpful imaginary exchanges. But you may be surprised how, once you start making them real in your mind, it's not long before people who are on your exact wavelength start calling you out of the blue, showing up on your doorstep, or crossing your path.

When I began this book, I knew the writing process would stretch me into new territory and that I would need guidance to find insights and understand new concepts. I made a point every day to imagine my soul group, and any spirit-writers interested in these topics, and ask them to join me in a collaborative writing process. When I'd get stuck, I'd invoke them by saying, "Hello everyone! What are we trying to say here? What are the right words?" And each time, the flow would restart from a more centered place.

Soon You Embrace Fellowship and Your Role in the Group Mind

By working with your soul group, you begin to practice fellowship; you *experience* the awe-inspiring cooperation among all souls. Really understanding fellowship means you no longer hold yourself as separate from the world and are open to influencing and being influenced by everyone—whether they're close, far, friend, foe, physical, or nonphysical. It means you *know* how other people's growth makes conditions easier for you and how your clarity helps them. Fellowship is based on mutual, conscious communion. You are your brother and sister's keeper, and they are your keeper. You learn that as you tend to others' needs, your needs are magically attended to, as well. The idea is to treat the other person as though they *are* you and to imagine in great detail how you would feel living in their body, seeing life through their eyes. Then imagine others stepping into your shoes to do the same.

Now you can feel how we all show up in each other's imaginations. Group experiences are created so the needs of everyone are met by the actions of everyone, where the cocreation of destinies is a most astonishing feat of spiritual engineering. When you witness the perfection demon-strated in nature by varied life forms in an ecosystem, for instance, you understand that what's needed is being supplied by the simple existence of all the other forms.

Try This!

Bless Someone or Something

When you realize the actual affect of directing positive thought and energy through your focused attention, you'll realize that "the art of bless-ing" is a powerful method for helping to heal people and things that are out of harmony. Call someone to mind, or think of the food you're about to eat at your next meal. Through empathic resonance, feel into the per-son or food and sense the lack of harmony. Then keep feeling further in, sensing the soul, the ideal inner blueprint. Hold your attention on that and let energy pour into it. Fill it up so you see the person or the food embodying its ideal. Know that this is the truth. Stay connected with the person or food until the vision feels like it takes hold and lives by itself. Remember: idealism comes from memory of the soul's reality and purity.

I recall Tom Peters, the management guru, describing what was really an experiment in fellowship. One company actually gave its secrets to its competitors, making it into a game so they could all "compete" better. The teams in their production facility helped each other, too, but competed at the same time, and in this way, everyone evolved to a higher level using a spirit of playful cocreativity. Fellowship means coming together with the idea that everything you do will further the other person somehow. With fellowship, the world wants you to win because then we all win. As our comfort level with fellowship increases, watch for new developments in our economic systems. It may start as various exchange and barter methodolo-gies and will progress into forms of outright sharing and philanthropy at a global level.

Working with the spirit of fellowship helps you learn what's possible to achieve by focusing the *group mind*—the merged awareness of all mem-bers of a group. If a group attunes its personal vibration to a single high frequency and asks questions or seeks creative insights for problem solving and innovation, the unique "filters" represented by the group members can precipitate a new kind of genius that's greater than the sum of the individual minds. By using imagination in groups, the group mind could invent a new product, imagine the design and testing process, work out the bugs, and see how it would sell in various markets and how long it would last. The group mind might also tune in to the product's timeline and see where snags might occur and why. Then, when the actual physical process comes to pass, it will materialize easily and quickly. For years, I've seen a recurring vision of teams of children gathered in a circle around a table, focusing their minds into a common frequency, drawing forth amazing ideas for futuristic technologies. I keep waiting to hear about it in the news.

Life is divine, life is an extraordinary, incredible, miraculous phenomenon, our most precious gift. We must grow a global brain, a global heart, and a global soul. That is our most pressing current evolutionary task.

Dr. Robert Muller

Try Communicating Through Time, Space, and Dimensions

You can learn to shift your frequency to a higher octave, to stretch beyond this world to a higher dimension and communicate with nonphysical beings. To do this, you need to realize that your imagination is simply a mental space in which different kinds of realities can be focused, like a staging area. Whatever you put your attention on, whatever you can think of, can occur in the space of your imagination. It's all experiential—not just three-dimensional physical reality but the full spectrum of higher-dimensional realities as well (see chapter 2). Your soul uses all the dimensions to create learning experiences that enhance your evolution. Dreams and meditation are the windows your everyday personality uses to peer into these very high-frequency, nonphysical experiences.

You might begin by centering yourself in your home frequency, in the moment, and just enjoy the quiet pleasure of your own being. Open the space of your imagination. Start to think about a personal council of spiritual teachers and advisors—which might include an intergalactic scientist or even an angel or two—that assists your evolution. Imagine them sitting at a round table, like King Arthur's knights, and there is a seat for you at that table as well. Imagine increasing your frequency, becoming 10 percent higher, then another 10 percent lighter, then another 10 percent more refined, until you've matched the frequency of the spiritual beings and can walk into the room and sit at the table with them. Then imagine a scene unfolding in which you ask questions, discuss ideas, learn something you need to know, or receive help with your energy flow.

Here's another scenario: picture someone you know who has died. See them in the meeting space of your imagination. As you do, they see you in their meeting space. Again, raise your frequency in increments, feeling yourself becoming lighter, brighter, and more transparent until you can walk into the meeting space with them. Look into their eyes. Feel their home frequency, their heart. You probably won't even have to speak because a direct telepathic transfer of thought will occur between you. Whatever needs to be communicated will be known. Or you might imagine connecting with a person from history who you admire and would like to talk to—perhaps Einstein, or Buddha, or Princess Diana.

> Oh they but mock us with a hollow lie,
> Who make this goodly land a vale of tears;
> For if the soul hath immortality,
> This is the infancy of deathless years.
>
> Alice Cary

By practicing these other-dimensional meetings with specific people, you can develop a wonderfully subtle and instantaneous sensitivity to anyone's personal vibration and what they know. Quick! Feel into Madame Marie Curie, Confucius, Napoleon, Katharine Hepburn, Jesus! This ability to feel into any person, idea, or group, whether physical or nonphysical, can help you tune to other highly specific things: for example, what it was like to live in London in 1705, or in Western China in 1254, or in South Africa when the first humans migrated to new parts of the world. You can learn to time travel and do your own firsthand historical research. You can modulate your personal frequency to match high priests in ancient Egypt, or Michelangelo's apprentices, or a pod of dolphins. You might frequency-match the vibration of particular minerals, plants, extinct animals, or kinds of music. Who knows what secrets might be unveiled to you? If you're toying with what might be possible to invent, you can harness these frequency-shifting ideas to find new sources of energy and forms of food, medicine and healing, art, transportation, environmental rejuvenation, and science.

What Does "Normal" Look Like in the Intuition Age?

In the Intuition Age, since you've learned to expand your identity to be more extensive than you ever thought possible, your level of common knowledge now includes new dimensions, which are really higher-frequency bands of awareness. You have tremendous, detailed information about both "outer" and "inner" space, which are seen to affect each other intimately because they aren't separate. I love the story of the German mystic and poet, Rainer Maria Rilke, who on a walk at night heard a bird singing in the stillness and felt the song coming from somewhere in the environment and from inside his body at the same time. You, too, now know this simultaneity and how the microcosm and macrocosm create each other and evolve together.

Separate disciplines, such as art, science, agriculture, religion, business, and government, have merged and become infinitely more sane and effective. You are adept in what were previously considered paranormal or supernatural consciousness skills, such as telepathy (direct transfer of thought between minds), teleportation (instant transfer of objects across time and space), clairvoyance and precognition (direct knowing across time and space), psychokinesis (movement of objects by intention alone), instantaneous spiritual healing, and the regeneration of physical structures from their inner blueprints.

When we have an experience of unlimited supply, possibility, and permission, creativity soars to new heights and the advancement of global culture proceeds at light speed. You see society transforming positively in a way that parallels your own growth, as many people are living their destinies simultaneously. Since you've become an expert in adjusting your frequency up and down the scales, it seems matter-of-fact to acknowledge beings from other dimensions, time periods, and parallel realities as kin and to work productively with them to create a better world. You teach and entertain yourself with your ability to travel in your imagination, as well as to materialize and dematerialize things in your world. You sense that a time is near when we won't experience death and birth; you and many others may enter and leave the world via ascension and descension. Many amazing things are in store for you in a transformed world.

Recognize the Mind's View and the Soul's View

A friend called the other day worried about things she'd heard in the news about politics, corporate greed, escalating prices, falling real estate values, and toxins in our food. She was swimming in a low-grade panic, having gone into the world's problems, their worst possible outcomes, and the potential negative effects on her family. Her energy, usually upbeat, was reminiscent of a deer frozen in headlights. I could almost feel her body trembling in response to the low vibrations she'd just matched on CNN. "What's going to happen?" she wanted to know. "Is everything going to be OK? What can we do?" I could relate, since I'd been struggling with the same thing and had for some reason just seen a string of "heavy" films about war-torn countries and the atrocities people are suffering in various parts of the world. I couldn't help thinking that what I'm writing about can seem incredibly naïve from one point of view. And yet from another point of view, it's absolutely real.

What I heard myself say to my friend was, "The only thing we can do now is stabilize our own personal vibration by not indulging in the seduction of fear, which is flooding the world and trying to saturate the air we breathe. It's like there are two kinds of air—dirty/toxic and transparent/clean—and they exist together in the same space. We have to choose to see the air as clear and nourishing, then breathe in that quality with every breath. The important work is to keep our bodies, emotions, and minds vibrating in harmony with our souls and spiritual values. Act as if there is nothing else. If the bad news tries to get us down, we don't let it. We don't ignore the news, but we don't let ourselves be flattened by it.

"At the same time, if ideas come to you about how to live more lightly on the planet, help those in need, or be creative while going with the flow of society's transformation process, do those things, but do them with joy and a spirit of generosity, not worry. Trade in the gas-guzzler for a saner vehicle, buy local produce, volunteer a few hours a week somewhere that appeals to you, vote for candidates who have higher consciousness. Do what you can do that feels good to you—we each have a form of self-expression that comes from our destiny that fits perfectly with what other people are doing. We don't have to sacrifice our happiness to do good in the world. If we just 'go for it' and stretch, fellowship will evolve the world. Not everyone is meant to be a political activist, yet at the same time, we've got to wake up in the places where we've allowed ourselves to be hypnotized into complacency by consumerism, advertising, government, and the media."

> As we are liberated from our own fear,
> our presence automatically liberates others.
>
> Nelson Mandela

After we spoke, I thought about it some more. There's the Mind's point of view and the Soul's point of view. The Mind looks at and for complexity, problems, and comfort through physical security and the familiar. This is the short view that reduces the idea of enlightenment to something ridiculous, sought only by unrealistic people. This view stays on the surface and makes the proliferating problems and chaos of the world into a *huge*, overpowering, hopeless situation that will most likely end in world destruction. It's hopeless, so why bother doing anything? Solutions, after all, are *so difficult to achieve* and *take such a long time*. This is the view that looks at all situations as black and white and either-or, and generates *truly endless* conflict between the right and wrong people. It tricks us by presenting powerful polarizing leaders who focus on enemies and crises and who absorb our attention. When we frequency-match this view, it looks so real! Who can doubt it? If you doubt it, it tells you you're naïve and stupid.

The Soul's point of view, on the other hand, is calm and looks at and for simplicity, unity, and compassion. It doesn't see the "terrible" situations on earth as problems, but as wave-turns and movements of one cycle to another, as symptoms of transformation. The Soul's view is long and deep;

it has the advantage of the clear overview, knowing there is no other possibility than for all beings to return to the Divine. It knows that solving "problems" *only* in the world of form is a waste of energy, that working with vibration and frequency can shortcut evolution and make transformation magical and nearly instantaneous. Nothing is hopeless, and enlightenment is close. The soul comfortably holds the paradox that nothing needs to change *and* there are things that are important to do. Movement into unity-awareness is inexorable, and the process is not linear and time consuming; it can happen any time you think to relax into it. The soul doesn't berate you when you don't choose its view; it waits patiently, smiling, inviting you back into the fold, into deep comfort.

Be a Love Teacher

I remember something John Denver said in his autobiography: "If not me, who then will lead? . . . If not me, who then?" You may not be satisfied with some aspect of your life, but there is always something you can do *right now* to improve things a little bit. You can make your own reality feel the way you want by choosing to feel your home frequency—your warmhearted, lovingkindness—in your body, emotions, and thoughts. You can end war in yourself. Much of the work of transformation and becoming transparent is simply about decluttering your personal energy-and-awareness field. By not giving attention to, and thereby dissolving, what's not vibrating at your home frequency, you can recognize yourself as the diamond light that remains. Think of it as shifting your attention from the Mind's complicated, defeatist view to the Soul's simple, empowered view. Even doing this is being a leader.

> Our awesome responsibility to ourselves, to our children, and to the future
> is to create ourselves in the image of goodness, because the
> future depends on the nobility of our imaginings.
>
> Barbara Grizzuti Harrison

For years, I've been interested in the Buddhist concept of *skillful perception*: healing your own pain and not adding more suffering to the world. I think leaders in the Intuition Age are going to be skillful perceivers and practical visionaries—those whose feet are firmly grounded on the earth, with heaven in every cell. You can be a leader, inventor, healer, or truth communicator. You can set an example for others by the way you live, how you share, and what you say. You can be a spokesperson for the soul. "Here's my heart, here's my mouth. Here are my hands and my feet. Move me where you want. Let me say the things that need to be said."

Instead of waiting for someone else to improve a situation, if you think of something that would help, you're the one who is supposed to do it. The collective consciousness caused you to notice the idea for a good reason. Don't be embarrassed to be idealistic; it's your highest vision vibrating through you. Your real work now, no matter the actual tasks you undertake each day, is to be a Love Teacher, Love Demonstrator, and Love Expander. You are creating yourself anew right now. You are helping Us by keeping your personal field vibrating at a high frequency and by seeing Us as part of you. And We are helping you in the exact same way. You and I are growing in such an amazing way that we may one day soon grow into what we can now barely even imagine; we may each become the consciousness of humanity—or even the universe—as a whole, and like the mystical, mythical unicorn in Rilke's sonnet below, our ideal Self may come into being simply because we have loved the potential and made space for it.

O, this was the animal that never was.
They did not know, but loved him anyway:
his smooth neck, graceful movements,
and the quiet light in his eyes.

True, he never was. But since they loved him,
a pure creature came to be. They made space for him.
And in this space, so clear and free and unbounded,
he easily lifted up his head and barely needed
what we call existence.

They nourished him, not with grain,
but always with the possibility of Being.

And this endowed the creature with such power
that a horn grew out of its forehead. One horn.
He went to a virgin, glistening white—and there,
inside a silver mirror and inside her, he was.

Rainer Maria Rilke, *Sonnets to Orpheus*, 2/4
translated from the German by Rod McDaniel

Just to Recap . . .

You are already close to a state of transparency, self-realization, or enlightenment, in which you easily recognize the difference between the Mind's reality and the Soul's reality and experience yourself as love and pure awareness. Transformation comes in staggered waves and is a path, not an endpoint. You may find yourself lightening up about the human condition while also becoming more compassionate. Self-realization in the Intuition Age is a combination of yang-masculine and yin-feminine ways of seeing spiritual growth. The path of Heart is the universal path that works for everyone. Empathy and compassion are the preferred modes of perception of the new heart-brain, and they teach you to care about life and feel commonality.

What then was the commencement of the whole matter? Existence that multiplied itself for the sheer delight of being and plunged into numberless trillions of forms so that it might find itself innumerably.

Sri Aurobindo

Empathic resonance is a healing force that overcomes isolation and helps develop safety, intimacy, and an easy grasp of truth. Compassion will surpass competition as the evolutionary methodology. You're learning to find and work with your soul group, the group mind, and the principles of fellowship to truly become part of the collective consciousness. In the Intuition Age, you'll be able to communicate with nonphysical beings and those who've died, as well as travel through time, space, and the dimensions. Right now, you can become a skillful perceiver, a practical visionary, and a Love Teacher.

Home Frequency Message

As I explain on page xxi in *To the Reader*, I've included these pieces of inspired writing at the end of each chapter as a way for you to shift from your normal, speedier reading mind to a deeper kind of direct experience. Through these messages, you can intentionally change your personal vibration.

The following message is meant to transport you into a way of knowing the world that's close to the way you'll experience life in the Intuition Age. To move into the *home frequency message*, just downshift to a slower, less hurried pace. Take a slow breath in, then out, and be as calm and still as possible. Let your mind be soft and receptive. Open your intuition and prepare to *feel into* the language. See if you can experience the deeper realities and feeling states that come alive *as you read*.

Your experience may take on greater dimension in direct proportion to the amount of attention you invest in the phrases. Focus on the words a few at a time, pause at the punctuation marks, and "be with" the intelligence delivering the message—live and right now—to you. You might speak the words aloud, or close your eyes and have someone else read them to you and see what effect they have on you.

ENTER THE SIMPLE LIFE OFTEN

You are on a long and winding journey, you are climbing the highest ladder, you are sailing the widest sea. And yet, you are standing still. Stars and planets rise and set, orbiting around you. The road moves beneath your feet, being pulled by a mysterious force, disappearing behind you. An ever-changing movie is being brought to you, a light show, an entertainment so stimulating and thrilling you can't take your eyes away. And you are creating it! Part of your ingenuity is pretending that you're not creating it, that it's coming magically from nowhere. How creative you are! And how mesmerized by your own brilliance.

Close your eyes now. Shut out your senses. Pull away from the stimulation and the changeable instability, and don't follow the waves traveling through time and space. They don't go anywhere that you aren't already. Feel your vibration shift as the outer fades—light and color dissolve into sound, which dissolves into the tiny tactile throb. Now that goes too, and there is stillness. In the quiet is a vibration that is so smooth it has no pulse, no discernable spaces between waves. This is peace. Be with it, and of its own accord, it sweetens. The state, within itself, smiles. That barely curving smile contains the entire love and wisdom of the Divine. It is the first sign of radiance, of soul-shine. Quietly radiating, and smiling because you can't help it. This is you. Yes, it is.

The simple life is the one most-real experience, common to all. It is not the kaleidoscope, not the pliable dream ever-melting and emerging. Real life is at the very bottom of any moment, and any moment from the movie can take you there. Feel it: the barely curving smile. Feel it: the slow, quiet shine. In that, knowing wafts around in you. Love is everywhere and unlimited. You don't need to possess any bit of it. You love to let it all drift, free.

Enter the simple life often and practice staying in it until your mind forgets what time is. The longer you stay, the easier it is to know it as your eternal permanent home. From here, smiling, you watch the kaleidoscopic movie tempting you to join it and invest in it fully. You love the movie, but you love the simple life, entirely. Everything here is clear. Knowing the simplicity, you master the frequencies. You love the particles and waves, you ride them and bless them as they move, as knowing moves, in you, without disturbing you, as the butterfly passes on a summer day, as the clouds dissolve and form on the high wind. You are everywhere and everywhere: it's simple.

Glossary

ascension: The ability to raise the frequency of one's body, emotions, and mind beyond the vibration of the physical world so the body disappears into a higher dimension without physical death. *Descension* is the reverse process of the body appearing in form without physical birth.

attunement: Adjusting the vibration of your body, emotions, and mind to match a particular frequency, usually of a higher vibration.

chakra: A vortex-like concentration of spinning, subtle energy located primarily along the spine; one of seven main centers of spiritual force in the etheric body.

clairaudience: The inner sense of hearing; to hear voices, music, and sounds without the aid of the physical ears.

clairsentience: The inner senses of touch, smell, and taste; to feel or sense nonphysical energy fields, discarnate entities, or patterns of knowledge without using the physical body also, obtaining information by touching objects.

clairvoyance: The inner sense of sight; the ability to see visions, past or future events, or information that can't be discerned naturally through the five material senses. Medical clairvoyance is the ability to view the human body and diagnose disease, while X-ray clairvoyance is the ability to see into distant or closed areas.

compassion: A pervasive understanding that knows love as the core of every being and situation. The virtue that gives rise to one's desire to alleviate another's suffering.

conscious communion: The act of intentionally merging with someone or something else, sharing a common experience, and experiencing intimate fellowship or rapport.

conscious sensitivity: Sensitivity applies both to physical and emotional feeling and is the strength of the sensation one experiences in comparison with the strength of a stimulus. Conscious sensitivity is the ability to immediately perceive subtle stimulation from either physical or nonphysical sources and to discern the meaning as it occurs.

crest/trough: The crest is a wave's high turning point; the trough is its low turning point.

destiny: Life after the soul has integrated fully and consciously into the body, emotions, and mind; one's highest-frequency life, accompanied by unlimited talents, harmonious energy flow, perfect timing, and doing what one is "built for" and most enjoys.

diamond light: A way to imagine the substance of the soul, diamond light conveys the experience of purity, supreme clarity and transparency, incorruptibility, and enlightenment.

dimensions: Levels or frequencies of awareness and energy progressing from physical to subtle or etheric, to emotional, to mental, to spiritual, and on into levels of the Divine. As awareness and energy progress through the dimensions, increasing frequency and greater unity is experienced.

direct experience: A live connection with the world, where one experiences situations immediately without pausing to analyze and compare; an experience of full engagement with each action in each moment.

Divine, the: A nonreligious way to refer to the Godhead, Creator, or Supreme Presence; an experience of perfect, transcendent force, truth, and love, or Oneness, with the universe.

$E = mc^2$: The formula describing Einstein's insight that matter and energy are different forms of the same thing. Energy (E) equals matter or mass (m) times the speed of light (c) squared. Einstein's formula reveals the amount of energy that mass would convert into.

ego: The sense of individuality based on separation from the whole. When awareness identifies as ego, fear and self-preservation are the motivators.

empathy: The ability to use one's sensitivity to feel "with" or "as" another person, group, or object, resulting in greater compassionate understanding.

enlightenment: The achievement of total clarity about the true nature of things and a permanent state of higher wisdom, illumination, or self-realization; the awakening of the mind to its divine identity; the final attainment on the spiritual path when the limited sense of "I" merges into supreme consciousness.

etheric or subtle energy: The vibratory frequency that is one level higher than matter; a malleable form of energy that acts as a kind of "modeling clay" or energetic blueprint for the physical dimension.

evolution: A gradual process of development, formation, or growth, especially one leading to a more advanced form; the growth of self-consciousness from the identity as a finite individual to unification with infinite, divine awareness.

feel into: The ability to penetrate into a person, object, or energy field with one's attention—to merge with it and become it briefly; to allow subtle information to register on one's body via conscious sensitivity, as if one is the object of observation.

felt sense: The impressions or direct experience of a person, object, or energy field registered on the body and mind through conscious sensitivity.

flow: The natural, continuous, fluid, wavelike movement of life and any process; a state in which one is fully immersed in thought or activity, characterized by a feeling of energized focus, full involvement, and enjoyment of the process.

fractal: A rough or fragmented geometric shape that is self-similar to its substructure at any level of refinement; it can be subdivided into parts, each of which is approximately a reduced-size copy of the whole.

frequency: The number of waves that pass through a specific point in a certain period of time; the rate of occurrence of anything.

frequency principles: The underlying dynamics of the way energy and consciousness function, especially after wisdom has been gained in the transformation process, which can be applied to improve the quality of daily life and the achievement of enlightenment.

frequency-match: The process of attuning one's personal vibration, whether consciously or unconsciously, to the vibration of another person for the purpose of communication.

frequency sorting: The process of weeding out people, situations, and opportunities with vibrations that don't match your home frequency, releasing them, and replacing them with people, situations, and opportunities that further your soul-expression.

harmony: A pleasing combination of the elements in a pattern that stresses the similarities and unity of all the parts.

hologram: A three-dimensional image (originally generated by a laser); a quantum mechanical explanation of reality that suggests the physical universe is a giant time-space hologram, the entirety of which is found within each facet, leading to the concept that every moment—past, present, and possible—exists simultaneously. Likewise, every place exists everywhere.

holographic perception: In contrast to *linear thinking*, the understanding that everything one notices is in the present moment, included in one's personal field, interconnected with, part of, aware of, and cooperative with everything else; sometimes called spherical perception because awareness feels like a radiating, expandable-contractible ball with one's individual self at the center; awareness that every particle contains the full awareness of every other particle, no matter the scale.

home frequency: The vibration of one's soul as it expresses through the body, emotions, and mind; a frequency of awareness and energy that conveys the most accurate experience of heaven on earth as possible.

inner blueprint: The most current pattern of one's life purpose or vision, the inner blueprint includes a mix of love and fear, wisdom and ignorance, and depends on how much personal growth and transformation a person has experienced. This energetic pattern gives rise to the events and forms in one's physical life.

Inner Perceiver: The power of the soul inside a person, sometimes known as the Revealer or Holy Spirit, that directs one's attention to notice things that aid in learning life lessons and expressing oneself authentically.

intuition: Direct knowing of what is real and appropriate in any situation without need for proof; perception that occurs when body, emotions, mind, and spirit are simultaneously active and integrated while being focused entirely in the present moment; a state of perceptual aliveness in which one feels intimately connected to all living things and experiences the cooperative, cocreative nature of life.

Intuition Age: The period following the Information Age, during which perception accelerates and intuition, direct knowing, and sensitivity take precedence over logic and willpower; the time on earth when soul awareness saturates the mind, transforming the nature of reality.

karma: The idea that the good and evil a person does returns either in this life or in a later one; a theory that whatever negative or positive energies one sends out come back to the sender in kind.

law of attraction: The idea of "like attracts like," that positive or negative thought and emotion attract positive or negative experiences in like proportion; one's thoughts determine one's experience.

law of correlation: The idea that because the inside and outside world aren't separate, if one has a thought, it is also occurring, or soon will, as an event in the world. Conversely, if one experiences a dramatic event, one holds a corresponding idea that brought the event into his or her awareness.

life-wave: The natural flow of one's life, moving from contraction to expansion, bringing ups and downs.

light body, or energy body: A higher-dimensional body that underlies or closely parallels the physical form, composed of etheric energy; often seen clairvoyantly as light. In it, the energies and functions of the chakras are coordinated and health and illness can be discerned as transparency, color, or shadows.

linear thinking: Perception characterized by cause-and-effect logic, the analysis of steps required to achieve a goal, and the idea that one must repeat what worked in the past to achieve the same results in the future.

many-worlds theory: The idea in physics that the world is split at the quantum level into an unlimited number of real worlds, unknown to each other, where a wave, instead of collapsing or condensing into a specific form, evolves, embracing all possibilities within it; therefore, all realities and outcomes exist simultaneously but do not interfere with each other.

materialization: The process of bringing an idea into physical manifestation. *Dematerialization* is the process of dissolving a physical form back into the unified field.

midbrain: the middle of the three primary divisions of the human brain that helps process the five senses, perceptions of similarity and connectedness, and affection.

misperception: A partial understanding or erroneous perception about the way energy, consciousness, and life function, leading to the development of an unhealthy feeling habit and needless suffering.

negative vibrations: Low-frequency or dissonant thoughts and emotions coming from oneself or others that cause one's personal vibration to drop.

neocortex: The last and most evolved level of the human triune brain, divided into a left and right hemisphere, involved in higher functions, such as spatial reasoning, conscious thought, pattern recognition, and language.

overlays: Unconscious, limiting beliefs one inherits in infancy and early childhood from parents and other influential people, which emphasize particular behaviors as necessary for survival.

past and parallel lives: The idea that souls are composed of thousands of parts, many incarnating into the earth to experience individual lives. The lifetimes of one soul can be separated from each other through time, giving the impression of sequential occurrence, or several lives may exist at one time but be separated by location.

personal field: The subtle energy around and through the physical body comprising an individual's pattern of etheric, emotional, mental, and spiritual energy and awareness; the radiance of one's personal vibration.

personal vibration: The overall vibration that radiates from a person in any given moment; a fluctuating frequency that is a combination of the various contracted or expanded states of one's body, emotions, and thoughts.

projection: Casting one's mind into thoughts of the past, future, fictitious realities, other locations, or other people's realities to avoid experiencing something in the present moment; blaming other people for what one doesn't want to acknowledge about oneself or seeing traits in others that one cannot see in oneself.

quantum mechanics: A branch of theoretical physics explaining the behavior of matter and energy at atomic and subatomic levels.

reptile brain: Also known as the reptilian complex, this is the first and oldest part of the human triune brain, concerned with instinct, emotion, motivation, and fight-or-flight survival behavior.

resonance: The vibration produced in an object due to the vibration of a nearby object; the regular vibration of an object as it responds to an external force of the same frequency. Waves that vibrate the same length create resonance. *Dissonance* is when vibrations of different wavelengths meet, create instability and chaos, and demand resolution.

shape-shift: The ability to make physical changes in oneself, such as alterations of age, gender, race, or general appearance, or changes between human form and that of an animal, plant, or inanimate object.

soul: The experience of the Divine expressing as individuality; the self-aware spiritual life force or essence, unique to a particular living being, carrying consciousness of all actions. The inner awareness in a person that exists before birth and that lives on after the physical body dies.

strike your tuning fork: The act of imagining that one's body and energy field are composed of the soul's frequency, then activating that resonance and imagining that the body, like a tuning fork, will radiate the vibration into everything it touches.

subconscious mind: Mental activity that functions just below the threshold of awareness, where all experience is stored as sensory data. The part of one's awareness where memories based on fear are stored or repressed.

synchronicity: A coincidence of events that seem to be related and which can be interpreted to find deeper meaning in life.

telepathy: The transfer of thoughts, feelings, or images directly from one mind to another without using the physical senses.

teleportation: The movement of objects or elementary particles from one place to another, almost instantaneously, without their traveling through space.

timbre: The special sound that differentiates one voice or instrument from others; a sound's tonal quality.

transformation: A complete change of physical form or substance into something entirely different; a total shift in the way reality functions.

transparency: A state of clarity and openness that is characterized by trust, spontaneity, and full engagement with the flow in any given moment; enlightened awareness.

truth and anxiety signals: The subtle, instinctive expansion or contraction responses of the emotions and body, sometimes experienced through the five senses, that indicate either safety and truth or danger and inappropriate action.

unhealthy feeling habits: Usually unconscious reactive behaviors caused by contracting and shutting off the experience of trust, love, and joy in early childhood or past lives; fear-based behaviors that are low frequency and which block the expression of the soul. As unhealthy feeling habits disappear, there is no need for any sort of feeling habit; one responds spontaneously to the stimuli that occur in each moment.

unified field: A universal sea of energy and awareness or presence that underlies and preexists physical matter; a state, force, or "ground of being" that is the absolute constant of the universe and connects everything in a single unified experience. Gravitational and electromagnetic fields, strong and weak atomic forces, and all other forces of nature, including time and space, are conditions of this state.

upscale solutions: Answers and decisions that resonate at the soul's frequency and which serve the growth of all people and the planet.

wavicle: A term devised to describe the dual nature of the behavior of the most elementary level of matter-energy in quantum physics. Matter and energy, at their core, can both act as a particle and a wave. Also known as a quantum entity.

yin and yang: The two equal and opposing forces in the universe, both necessary for harmony. Yang energy and awareness is external, assertive, creative, warm, dry, bright, and male. Yin is internal, receptive, dissolving, cool, wet, dark, and female. Each contains a small amount of the other.

Free Audio Companions for
Frequency: The Power of Personal Vibration

Visit www.beyondword.com/penneypeirce/ to download five free audio companions designed and narrated by Penney Peirce to further enhance your work with your personal vibration and home frequency.

Track 1: Introductory Message

Penney gives you some background about what's happening today in the realm of energy, explaining how your personal vibration is changing, that there are stages to the transformation process, and you *can* learn to navigate this process easily and quickly. She introduces four meditations and talks about how they can be experienced repeatedly and what you can do to receive maximum benefit from them every time you listen. (approx. 7 minutes)

Track 2: Meditation 1

Be Conscious of Your Frequency: Track Your Daily Vibrations
It's useful to develop a mindfulness habit that focuses on how high or low your vibrations are during the day, so you can recognize and consciously choose to remain in your home frequency, or "preferred state." In this meditation, Penney helps you focus on and track the differences between your home frequency and the lower energy states of people and places you may have unconsciously matched throughout the day. (approx. 19 minutes)

Track 3: Meditation 2

Experience Your Home Frequency: Focus on Positive Qualities and Memories
In this two-part meditation, Penney guides you in activating your home frequency by first helping you frequency-match various positive states of being that allow your soul to flow through you freely. In part 2, she helps you reintroduce memories of positive experiences you've had to your body and emotions, letting them become real again, building a composite of qualities that open and maintain your home frequency state. (approx. 21 minutes)

Track 4: Meditation 3

Experience Your Home Frequency: Activate Your Diamond Light Body
Penney talks about the power of using diamond light as a meditation image and guides you into an experience of it. Then she helps you bring your Diamond Light Body—the energy body of your soul—inside yourself and into perfect resonance with all the parts of your body. As you experience this, you become your Diamond Light Self and realize you really do know what you're doing. You see how you're in direct contact with the unified field and collective consciousness. (approx. 13 minutes)

Track 5: Meditation 4

Moving Smoothly Through the Vibrations: Penney Reads the Frequency Message from the End of Chapter 2
This piece of writing was generated in a high-frequency, home-frequency state. By slowing down and tuning in to the energy in the words and phrases, you'll be able to frequency-match this state. (approx. 6 minutes)

Listen to these meditations when you can close your eyes, truly relax, and not endanger yourself or others. Do not listen to these recordings while driving a vehicle or operating machinery.

Running time approximately 66 minutes / ℗ & © 2011 by Penney Peirce Communications.

Find out more about Penney, her books, and audio-video interviews at: www.penneypeirce.com and www.thefrequencybook.com.